Particle Physics in the Cosmos

Particle Physics in the Cosmos

. . .

READINGS FROM
SCIENTIFIC AMERICAN MAGAZINE

Edited by

Richard A. Carrigan, Jr.
Fermi National Accelerator Laboratory

and

W. Peter Trower
Virginia Polytechnic Institute and State University

W. H. FREEMAN AND COMPANY
New York

Some of the SCIENTIFIC AMERICAN articles in *Particle Physics in the Cosmos* are available as separate Offprints. For a complete list of articles now available as Offprints, write to W. H. Freeman and Company, 41 Madison Avenue, New York, New York 10010.

Library of Congress Cataloging-in-Publication Data

Particle physics in the cosmos : a Scientific American reader / edited
    by Richard A. Carrigan and W. Peter Trower.
        p.   cm.
    Bibliography: p.
    Includes index.
    ISBN 0-7167-1919-3
        1. Nuclear astrophysics.   2. Particles (Nuclear physics)
3. Cosmology.   I. Carrigan, R. A.   II. Trower, W. Peter.
III. Scientific American
QB464.P37   1989   88-32220
523.01′972 — dc19   CIP

Printed in the United States of America

1 2 3 4 5 6 7 8 9 0 RRD 7 6 5 4 3 2 1 0 8 9

CONTENTS

SECTION V: PARTICLES AND THE UNIVERSE

Note on cross-references to SCIENTIFIC AMERICAN *articles*: Articles included in this book are referred to by chapter number and title; articles not included in this book but available as Offprints are referred to by title, date of publication, and Offprint number; articles not in this book and not available as Offprints are referred to by title and date of publication.

Introduction

The past two decades have seen far-reaching breakthroughs in our understanding of fundamental physics. Through these breakthroughs we can now explain much about the tiniest particles as well as the mighty universe itself. More amazing still, these developments on staggeringly different scales now appear to be inexorably linked.

That protons and mesons are formed from fractionally charged quarks has gained universal acceptance. These quarks seem to be confined to their particles by colored gluons. Two of the forces that govern quark behavior — electromagnetism and weak interactions or radioactivity — have been neatly joined as one. So-called Grand Unification Theories offer hope for further unification with the strong nuclear forces. All of this understanding has resulted from research with powerful accelerators and increasingly sophisticated experimental detectors.

On the cosmic scale the discovery of the three-degree black-body radiation has led us to believe that the universe is remarkably uniform or homogeneous. Further, protons are rare compared with these three-degree photons that pervade the universe. The discovery in 1964 of the violation of previously sacrosanct time reversal symmetry (strictly CP violation) has provided a mechanism to explain this ratio and the absence of antimatter in the universe. However, in the late 1970's a fly was found in the ointment — the Grand Unification Theories (known as GUTs) predicted many heavy magnetic monopoles but few if any now seem to be there. A new inflationary cosmology has since solved the problem of too many monopoles and explained the large-scale uniformity of matter in the universe. Thus the early moments of the universe — the birth of matter as we know it, the world as we know it, the cosmos as we know it — appear to tightly link the biggest bang with the smallest particle.

During the past two decades SCIENTIFIC AMERICAN has skillfully presented these original developments, using the best possible reporters — the scientists who were creating these developments. In this volume a selection of SCIENTIFIC AMERICAN articles chronicles the most recent developments in fundamental physics and cosmology. Some of the authors are among the most distinguished physicists and astronomers in the world today.

Particle Physics in the Cosmos ties the physics of elementary particles to the universe itself. Lawrence

Krauss begins with a discussion of the now famous dark-matter problem. The very structure of the universe and how physical theories can influence this structure is next. Joseph Silk and John Barrow tell us about an idea for what the universe was like soon after its birth. Then Silk, with his Soviet colleague Yakov Zel'dovich and Alexander Szalay of Hungary, speculates about how the large-scale objects in the universe evolved.

The idea that all the known forces of nature were unified in one basic force in the beginning of the life of the universe is appealing. Howard Georgi tells us that all is not lost for the simplest proposal to unify the electroweak and strong interactions. Then Gerard 't Hooft talks about a class of theories based on symmetries in nature that in principle predict the complete pantheon of particles, discovered and undiscovered.

However attractive the idea of unifying the forces of nature is, that it actually happens must be demonstrated by experiment. The hope for an unstable proton is expressed by theorist Steve Weinberg. Experimentalists John LoSecco, Fredrick Reines, and Daniel Sinclair then dash this hope by concluding that as far as they can tell so far the proton does not decay. Next, Richard Carrigan and Peter Trower review the case for the exotic magnetic monopole spawned by these unification theories.

In the last section of the book elementary particles and their role in the universe are discussed. Jay Pasachoff and William Fowler tell how the isotope of hydrogen (deuterium) alone among the elements may tell us the density of nuclear matter in the universe. Frank Wilczek discusses how antimatter fits into the ideas of the nature of the universe. How we got from the big bang to where we are now with particle abundances we see today is described by Alan Guth and Paul Steinhardt. Finally, Duane Dicus, John Letaw and Doris and Vigdor Teplitz speculate about how this universe of ours will end.

A reprint volume is special by its nature. Each article was written independently to be self-contained, so there is bound to be repetition in the collection. Likewise there are a few omissions (e.g., three-degree radiation, supersymmetry, abundance of light elements, etc.). The repetition offers advantages. First is timeliness—it would take years for one or two authors to pull together material at this level. Second, each author has a different point of view so that the reader can get an individual perspective from each article. The third advantage is that if one article seems too long or too abstract, the reader can try a different one. Sometimes it helps to read the experimental articles first for an easier view of theory. But be warned—some of the longest and most difficult articles also offer some of the greatest insights into the underlying physics.

For whom is this book intended? We have addressed four audiences: serious high-school students, university students, intelligent general readers, and scientists themselves. Some of the articles are not simple, but they offer the serious and intelligent reader a unique insight into the thinking of the scientists that did the work.

An important feature of these articles is the guiding hands of the SCIENTIFIC AMERICAN editors and their skillful illustrators. The editor plays an important role in bringing the involved scientist down to traditional language. Most of the authors have felt that the editors drove them too far toward popularization while their editors believed the scientists could never appreciate the need for clarification. The result is a compromise.

An ideal use for the book would be to give it to a superior high-school student and ask him or her to guess what will happen next. Our present picture of physics and cosmology is incomplete and some things are undoubtedly wrong. This is the fabric of science, but as Haim Harari has said, "Imagining what a successful model might be like, however, is not at all the same thing as actually constructing a realistic and internally consistent one" (see "The Structure of Quarks and Leptons," SCIENTIFIC AMERICAN, April, 1983).

Another word of warning. Particle physicists have recently become enamored with words beginning with "s," with "standard" and "super" being the worst offenders. In this volume the standard model in one section may refer to cosmology, in another to electroweak unification, and in a third to some form of grand unified theory.

To help with the interrelating and updating of these articles we have added postscripts, often by the authors themselves. These are minor emendations since most of the articles have stood the test of time remarkably well.

The articles were drawn from more than twice the number that appeared on the subject in SCIENTIFIC AMERICAN. Space limitations have forced us to omit some important articles. We have tried to balance experiment-theory, particle-cosmology, our own prejudices and internal discord and other factors. To our friends and colleagues whose articles do not appear here we can only say that the other of us, a most obdurate and unthinking person, was responsible.

We wish to acknowledge those who have contributed to this project. The busy authors have responded by adding one more chore to their already overfull schedules. The editors of SCIENTIFIC AMERI-CAN have played an essential role by selecting and shaping the original articles. Rita Gold and Louise Ketz have skillfully managed the transition from the original magazine to the brand-new format.

Richard A. Carrigan, Jr.
W. Peter Trower

SECTION

THE COSMIC CONNECTION

. . .

Dark Matter in the Universe

More matter exists than is seen. The motions of stars and galaxies indicate where some of it is; theory suggests there is far more. What and where is it? Particle physics and astrophysics are yielding clues.

• • •

Lawrence M. Krauss

What is the universe made of? What kind of matter is commonest, how much is there and how is it distributed? These questions, always a focus of cosmology, have become even more intriguing over the past few years as evidence has piled up to support the proposition that most of the mass in the universe is dark — invisible to any existing telescope or other observational device — and new developments in both high-energy physics and astrophysics have made possible new predictions of the makeup and distribution of this possibly exotic form of matter.

There is already overwhelming evidence that the visible matter within galaxies may account for less than 10 percent of the galaxies' actual mass: the rest, not yet directly detectable by observers on the earth, is probably distributed within and around each galaxy. Theoretical considerations now suggest this may be only the tip of the cosmic "iceberg" of dark matter: much greater amounts of dark matter may be distributed throughout the universe, perhaps in configurations entirely independent of the distribution of galaxies. It may be that this mass can be accounted for only by the existence of new kinds of matter.

The question of dark matter — how much of it there is, how it is distributed and what it is made of — is intimately linked to questions about the overall structure and evolution of the universe: because dark matter is probably the dominant form of mass in the universe, it must have affected the evolution of the features observable today. Questions of structure in turn depend for their answers on a deep bond that has formed between macrophysics and microphysics, the bodies of knowledge that respectively describe interactions on the largest scale (that of the universe as a whole) and the smallest scale (that of the fundamental particles that make up all matter).

This bond is provided by the observation that the universe is expanding. If we are bold enough to extrapolate the expansion backward by between 10 and 20 billion years, the cosmological and microscopic scales begin to merge, because at the earliest times those structures now observed on the largest scales occupied regions having characteristic distances and energies on scales that are typically associated with the processes governing the interactions of fundamental particles. Since the structure remaining on the largest scales observable today reflects the imprint of those processes, it is natural to expect the resolution of the dark-matter question to come in part from advances in the understanding of the physics of high-energy particles.

At present a number of testable predictions for the nature of both the dark matter and the primordial structures in the early universe have been proposed. Future developments, both theoretical and observational, will help to decide issues ranging from how and when galaxies and stars first formed to what kinds of symmetries underlie the interactions of particles at very high energies. Ultimately the debate about dark matter may help to answer a question as old as human inquiry: What will be the fate of the universe?

Ever since the early 1930's, when Edwin P. Hubble confirmed that the universe is expanding, it has been natural to ask whether the expansion will eventually halt. The answer depends on two factors: how fast the universe is currently expanding and how strongly the force of gravity, determined by the average density of mass within the universe, holds that mass together. A high mass density would cause a strong gravitational attraction.

According to the general theory of relativity, there is a relation between the magnitudes of these two factors and the mean curvature of the universe. If the average mass density is so small compared with the expansion rate that the universe will continue to expand at a finite rate forever, the universe is said to be open. If the density is high enough to halt the expansion and cause the universe to contract again, the universe is said to be closed. If the gravitational attraction is precisely strong enough to continue to slow the expansion but not strong enough to close the universe, the universe is said to be flat. The shape of space will affect the shape of geometric objects and the reference length, which is the difference between any two regions of the expanding universe (see Figure 1). Over small distances, such as on the earth, these measures would not be noticeable. Strong theoretical arguments support the proposition that the universe is actually flat, even though in order to be flat it would have to contain much more mass than has yet been observed, either directly or indirectly.

Because the observable universe is highly uniform in all directions, its rate of expansion can be described in terms of a single parameter, which is known as the Hubble constant even though it is actually a slowly varying function of time. The Hubble constant is the average speed with which any two regions of the universe are moving apart from each other divided by the distance between them (see Chapter 2, Figure 7).

For any given measurement of the Hubble constant, it is easy to determine the mass density that would correspond to a flat universe. Measurements of the Hubble constant, however, depend on a variety of uncertain measurements. The Hubble constant is generally determined by measuring the velocity at which various objects are receding from the earth and gauging their distance by such techniques as estimating their intrinsic brightness and comparing that with their brightness as seen from the earth.

Because those measurements are highly uncertain, there is a spread of about a factor of two in current determinations of the universe's rate of expansion. As an upper limit, objects one megaparsec (about 3.26 million light-years) apart are on the average receding from one another at a speed somewhat less than about 100 kilometers per second. At that rate the average mass density that would result in a flat universe is about 2×10^{-29} gram per cubic centimeter, which is roughly equivalent to the mass of 10 hydrogen atoms per cubic meter of space.

How is it possible to determine how much mass actually exists? One method for finding at least a lower limit is simply to add up the total amount of visible matter. Since what can be measured directly is not mass but luminosity, some amount of interpretation is necessary in translating observations into putative mass densities. When the observed distribution and luminosity of stellar objects and diffuse gas are taken in combination with theoretical estimates of their masses, it seems that the mass-to-luminosity ratio of the luminous matter associated with galaxies is a few times the mass-to-luminosity ratio of the sun. Given this estimate and estimated lower limits on the Hubble constant, the average density of luminous matter in the universe is less than about 2 percent of the density needed to halt the universe's expansion.

It has been known since as early as 1933, however, that clusters of galaxies may contain a significant proportion of nonluminous mass. In that year Fritz Zwicky of the California Institute of Technology was analyzing the individual velocities of galaxies within the Coma cluster. He found many galaxies were moving so quickly that the cluster as a whole should tend to fly apart unless there was more mass to hold it together than the luminous mass alone. Other evidence indicated the cluster was stable, and so Zwicky concluded that the cluster must contain nonluminous matter.

Zwicky set an important precedent by showing

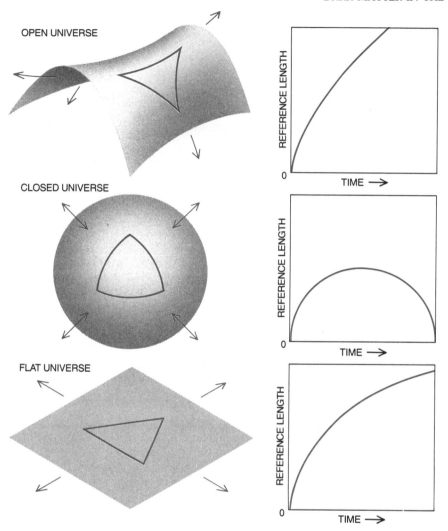

Figure 1 CURVATURE OF THE UNIVERSE. In an open universe geometry is analogous to the surface of a saddle, a triangle has an angle-sum of less than 180 degrees and a reference length between regions continues to increase. A closed universe closes in much the same way as the surface of a sphere, the angles of a triangle add up to more than 180 degrees and a reference length eventually begins to decrease. In a flat universe geometry is analogous to a plane, the sum of the angles in a triangle is exactly 180 degrees and expansion continues but slows asymptotically.

that dark matter can in principle be detected indirectly by its gravitational effects. In recent years investigators have shown convincingly that similar techniques can detect the presence of dark matter in structures on scales ranging from the immediate solar neighborhood through galaxies and clusters of galaxies to superclusters made up of thousands of galaxies.

The best-documented evidence for the presence of dark matter is based on the velocities of rotation of spiral galaxies [see "Dark Matter in Spiral Galaxies," by Vera C. Rubin; SCIENTIFIC AMERICAN, June, 1983]. The Doppler frequency shift makes it possible to determine how quickly a light-emitting object is moving toward or away from an observer and how fast the arms of a spiral galaxy are rotating. A stellar object emits light at characteristic frequencies determined by its composition. If the object is mov-

Figure 2 SPIRAL GALAXY M31 (ANDROMEDA) reveals the presence of dark matter by the motion of its outer arms, which rotate about the galactic center faster than they would be expected to if the galaxy's visible, luminous matter represented most of its mass. (Palomar Observatory Photograph.)

ing away from the observer, the wavelength of the observed light appears to be lengthened. This is called a red shift, because longer-wavelength light is redder. Nearly all galaxies are moving away from the earth because the universe is expanding. To an observer on earth the wavelength of the light from the spiral galaxy therefore appears to be lengthened, or red-shifted (see Figure 3). By comparing the red shifts of the galactic center and the arms, the rate of rotation of any part of either arm can be determined. It is then possible to infer the distribution of mass in the galaxy.

The velocity of rotation of an object in a stable, gravitationally bound system, such as a spiral galaxy, depends in part on its distance from the center of rotation. According to Newton's laws, the orbital velocity of objects far from a central concentration of mass should drop off in proportion to the reciprocal of the square root of their distance from the center of rotation. In extensive surveys of stars and hot gas in the outer regions of spiral galaxies, several groups have shown that the rotational velocities

of these objects remain constant, rather than dropping off, out to distances greater than 30 kiloparsecs from the galactic core. It had already been suggested by Jeremiah P. Ostriker and P. James E. Peebles of Princeton University that there must be some unseen mass in spiral galaxies, because otherwise gravitational instabilities would cause the galaxies to collapse into barshaped formations. The stability of spiral galaxies, as well as the rates of rotation of their outer arms, could be explained if the galaxies were each embedded in a large, roughly spherical distribution of dark matter.

There is other dynamical evidence for dark matter, on scales both larger and smaller than the scale of individual galaxies. The evidence is obtained not from measurements of rotational velocities but from measurements of the random individual velocities of objects within gravitationally bound systems. A well-known theorem of classical mechanics called the virial theorem establishes a

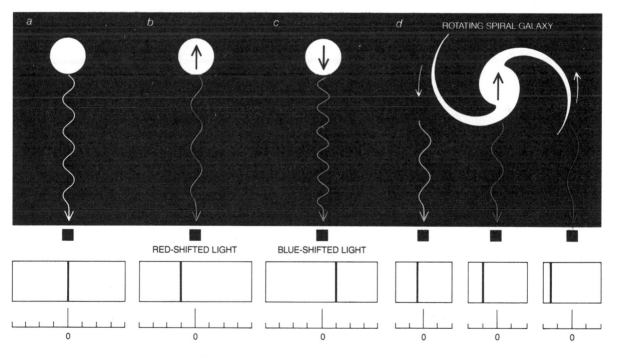

Figure 3 DOPPLER FREQUENCY SHIFT. A stellar object emits light at frequencies determined by its composition (a). If the object is moving away from the observer (b), the wavelength of the observed light is lengthened, or red-shifted. If the object is moving toward the observer (c), the wavelength of the light is shortened, or blue-shifted. The light from the center of a galaxy that is moving away from the earth is red-shifted (d, center). One arm of the spinning galaxy (d, left) will not be moving away from the earth as quickly as the galactic center, so its light will be less red-shifted. The other arm will be moving away more quickly (d, right), so its light will be even more red-shifted.

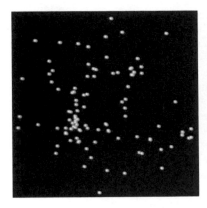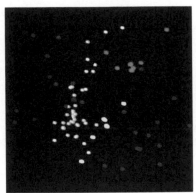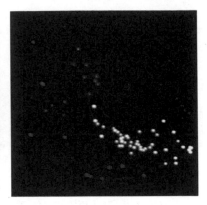

Figure 4 CLUSTER OF GALAXIES IN CANCER, as these computer-generated views by Michael J. Kurtz show, is not a single dynamical system. From the earth (*left*) the cluster seems to be a roughly spherical system in apparent equilibrium. Analysis showed that the cluster is made up of several groups of galaxies separated in space (*center*, colors). A rotated view (*right*) shows the separation of the various groups more clearly. Within each group the relative velocities of galaxies are much lower than the relative velocities of the group, indicating there is less mass in the system as a whole than previously estimated.

relation between the average kinetic and gravitational potential energies of objects in stable, gravitationally bound systems that have reached dynamical equilibrium. It should therefore be possible to estimate the total mass of such a system (which is related to its total gravitational potential energy) by measuring the relative velocities of a large number of pairs of objects within the system. This method has yielded evidence of dark matter in a wide variety of systems, ranging from dwarf spheroidal galaxies as small as 10^7 solar masses to clusters of galaxies as large as 10^{15} solar masses. On the largest scales probed by this kind of analysis (regions within roughly a megaparsec of galaxies) the average mass densities are no larger than about 20 percent of the density needed to close the universe.

Another method, pioneered by Peebles and his co-workers, relies on statistical analysis of large numbers of galaxies rather than on data taken from individual galaxies or clusters. Peebles showed that by amassing statistical data on galactic motion and clustering on different size scales it is possible, under the assumption that the regions probed contain gravitationally stable dynamical systems, to relate the mean relative velocity of a large number of pairs of galaxies to the mean mass density of the universe.

It is striking that all the available methods, including those I have discussed and several I have not mentioned, yield essentially the same result: if the distribution of galaxies traces the distribution of mass in the universe, then the universe contains less than about 20 to 30 percent of the mean mass density that would be necessary for closure.

Even if galaxies are not good tracers of mass, or if somehow all the analyses have involved systematic errors, there is still good reason to believe that at any rate the total amount of ordinary mass (mass consisting mainly of protons and neutrons) in the universe accounts for no more than about 20 percent of the amount that would be required for closure. The evidence comes for the most part from the theoretical framework that explains the process of nucleosynthesis, in which various cosmically abundant light elements and isotopes were first formed.

Nucleosynthesis of light elements occurred primarily in the first few minutes of the universe's existence. The process of nucleosynthesis would have been extremely sensitive to the absolute density of protons and neutrons at that time. In order for the predictions of current theoretical models of nucleosynthesis to agree with the present-day abundances of the light elements, the total density of protons and neutrons that could have been present at the time of nucleosynthesis is constrained so tightly that these particles' current density must be less than about 20 percent of the density required for closure. Thus it seems that if the universe is closed, at least 80 percent of the total mass in it is made up of some other kind of matter.

Since such fundamental theoretical arguments limit the amount of normal mass in the universe to 20 percent of the critical density, and since observational evidence suggests that the mass density associated with galaxies and clusters of galaxies is about that amount, why should cosmologists not assume the universe is in fact open? It is by no means impossible to imagine a form in which enough normal matter to explain the dynamics of galaxies and clusters could remain unseen. Why, then, is there a need to postulate any other form of mass? Why is there a larger dark-matter problem?

Two theoretical barriers stand in the way of the simple assumption that most or all of the mass in the universe is composed of normal matter and that the mean density is only 20 percent of the critical amount. The first barrier is set by a combination of the theory of galaxy formation and observations of the background of microwave radiation that pervades the cosmos.

It is generally assumed that galaxies eventually formed when regions of the early universe that were denser than the average condensed under the force of gravity until they separated from the background expansion to form isolated bound systems. For roughly 100,000 years after the big bang, ordinary matter could not condense in this way. Ordinary matter was still too hot for its constituent particles to have combined into electrically neutral atoms, and so it consisted of independent charged particles. Because ordinary matter was ionized in this way, its microscopic motion was strongly influenced by background fields of electromagnetic radiation: matter and radiation were coupled. Regions of ordinary matter that were denser than surrounding regions and smaller than the horizon size (the distance a light ray could have traveled since the big bang, and therefore the maximum distance over which physical systems could be in causal contact; see Chapter 3, Figure 18) could not have condensed further, because the "pressure" of the radiation combated the attracting force of gravity.

Eventually the universe had cooled enough for oppositely charged particles to combine, rendering normal matter electrically neutral, and so matter decoupled from radiation. The thermal background-radiation bath to which the matter had been coupled was then free to cool as the universe expanded, and it now constitutes the well-known cosmic microwave background radiation, which fills the universe. Observations have shown that this background radiation is isotropic—the same in all directions—to within a very high degree of accuracy.

Since gravity is a universally attractive force, any initial fluctuations, or small variations, in the density of ordinary matter in the early universe would have tended to grow after the force of radiation pressure no longer acted against the force of gravity. Thus it is presumed that the universe became (and is becoming) clumpier with time and that galaxies, whose cores now have densities more than one million times the average background density, began in fluctuations whose densities were much closer to the background value.

How large were the initial fluctuations? Because of the limited data currently available on large-scale structures, and because of the mathematical difficulties inherent in describing analytically the evolution of systems as dense as galaxies, it is extremely difficult to work backward from the current state of the universe to determine the precise nature of the initial fluctuations. An easier approach is to assume some initial pattern of fluctuations, simulate the growth and evolution of that pattern and compare the result with present-day observations. In this approach the cosmologist is guided by both lower and upper limits on the size and nature of the initial fluctuations. First, they must have been extreme enough (that is, the ratio between the local overdensity in the region of the fluctuation and the average density in space must have been large enough) for fluctuations on the scale corresponding to galactic sizes to have condensed to form galaxies by today. Second, the fluctuations must have been of small enough amplitude for them not to have left an anisotropy in the background radiation larger than the measured upper limit.

These two conditions appear to be mutually inconsistent if the universe is composed mainly of normal matter. Between the time when normal matter became decoupled from radiation and the time when the fluctuations that would become galaxies collapsed to form isolated, gravitationally bound systems, the initially small fluctuations in density could grow only at a well-defined rate. Fluctuations large enough to have had sufficient time to form self-bound systems would have led to an anisotropy on the background radiation more than an order of magnitude greater than the observational upper bounds. In other words, there has not been enough time, since decoupling, for galaxies to form gravitationally from variations in density small enough not to have left observable traces in the background radiation.

This conclusion depends on two widely held assumptions, namely that the microwave background has not been significantly disturbed since the time of decoupling and that gravity alone led to the formation of galaxies. Unless either of these standard assumptions is false (as various investigators have suggested), it appears that some new form of matter is necessary, one that could have begun to condense gravitationally earlier than normal matter could have.

There is a second and more fundamental reason to suppose the universe is not dominated by normal matter having a density of only about 20 percent of the critical density. This reason, now called the flatness problem, was first pointed out by R. H. Dicke of Princeton and Peebles. The essential point is that any deviation from an exactly flat universe should tend to increase linearly with time. If the universe had had even a small nonzero curvature at the time of nucleosynthesis, the deviation from flatness would by today have increased by a factor of about 10^{12}. Since the mass density in the present-day universe is within a factor of 10 of the mass density of a closed universe (in other words, since the universe is relatively close to being flat), at nucleosynthesis the universe must have been either exactly flat or curved to an extremely small degree: it must have been flat to an accuracy of within one part in a million million.

If the universe is measurably curved today, cosmologists must accept the miraculous fact that this is so for the first time in the 10^{10}-year history of the universe; if it had been measurably nonflat at much earlier times, it would be much more obviously curved today than it is. This line of reasoning suggests that the observable universe is essentially exactly flat: that it contains precisely the critical density of mass. Since normal matter probably accounts for only 20 to 30 percent of the critical density, some form of more exotic matter is probably present.

The next logical question is: Why is the universe exactly flat? In 1980 Alan H. Guth, now at the Massachusetts Institute of Technology, proposed an answer. It took the form of a model of the evolution of the early universe based on ideas in particle physics that had only recently been proposed.

Guth drew on the work of Howard Georgi and Sheldon Lee Glashow of Harvard University. In 1974 the two investigators proposed that three of the fundamental forces of nature—the so-called strong, weak and electromagnetic forces—are different aspects of a single, "unified" force. At sufficiently high energies the three forces should be exactly symmetrical: they should behave identically. At energies comparable to those observed now on the earth, on the other hand, the three forces can behave quite differently (see Chapter 4, "A Unified Theory of Elementary particles and Forces," by Howard Georgi). The temperature of the early universe, soon after the big bang, was initially high enough for the symmetry of the three forces to be manifest. As the universe cooled below the critical energy at which the symmetries relating the forces can be maintained, the preferred configuration of the universe became one in which the symmetry was "broken." The effect of this symmetry breaking was that the forces appeared distinct from one another.

(A simple example of this type of behavior is found in ferromagnets. At sufficiently high temperatures a piece of iron is not magnetized: the spins of all the electrons, each of which causes a small magnetic field, point in random, different directions. Below a certain critical temperature, however, it may be energetically more favorable for all the spins to point in one direction, aligning their magnetic fields and creating a permanent magnet. The direction of the magnetic field in the magnet represents a unique direction, and so the symmetry of the former configuration, in which no direction was special, is broken.)

According to Guth's idea, which was later extended by Andrei D. Linde of the P. N. Lebedev Physical Institute in Moscow and by Paul J. Steinhardt and Andreas Albrecht of the University of Pennsylvania, the abrupt breaking of symmetry could have caused the universe to "inflate" rapidly: the universe could have expanded exponentially, growing by more than 28 orders of magnitude in less than 10^{-30} second. After the period of rapid inflation the universe could have reverted to its normal, nonexponential expansion, which is observed today (see Chapter 11, "The Inflationary Universe," by Alan H. Guth and Paul J. Steinhardt).

It is the rapid inflation of the universe, according to this model, that caused the observable regions of space to become flat, in much the same way as inflating a balloon makes its surface appear flatter; after inflation the part of the universe observed today would necessarily appear flat.

In addition to its resolution of the flatness problem, the inflationary-universe scenario is remarkably successful in other ways. In particular, it is the only model consistently tying the initial conditions that caused the universe's expansion to the laws of microphysics. The inflationary model also makes it possible to calculate, from first principles, quantities whose values had previously been assumed or inferred. For example, the model remarkably predicts the shape of the spectrum of primordial density fluctuations (the functional relation between the amplitude of fluctuations and their scale size) to be precisely the shape that had been suggested earlier on phenomenological grounds. The wide acceptance by many cosmologists of the predictions of the inflationary-universe model indicates the deep impact particle theory is having on modern cosmology.

In solving the flatness problem, the inflationary model makes the dark-matter problem more urgent. If the universe is flat, then most of the mass in the universe is probably not normal matter, and most of it has not yet been detected in any way, even indirectly.

What might this exotic, undetected matter be made of? One of the earliest proposals was that the dark matter is composed of neutrinos. First postulated in order to solve problems involving the conservation of energy and momentum in nuclear decay, neutrinos interact very weakly with normal matter and are thus extremely difficult to detect. Nevertheless, three kinds of neutrino, called the electron neutrino, the muon neutrino and the tau neutrino, have now been found experimentally. It was originally proposed that neutrinos were massless, but there is no theoretical reason for supposing they might not have some mass. Stringent experimental limits have nonetheless been set on the maximum possible neutrino mass, and it is very small indeed. The strongest constraint is on the electron neutrino, which must have a mass less than about 10,000 times smaller than the mass of the electron.

As dark-matter candidates, neutrinos have two strong advantages over other contenders. First of all, they are known to exist. Second, the calculations that have been so successful in describing primordial nucleosynthesis also suggest that light neutrinos must be abundant in the universe today. When big-bang nucleosynthesis started, at temperatures greater than 10^{10} degrees Kelvin (degrees Celsius above absolute zero), light neutrinos were kept in thermal equilibrium with matter by the weak interaction and were therefore as abundant as photons. Thus, as R. Cowsik of the Tata Institute of Fundamental Research in India and J. McLelland of the University of Melbourne first estimated, if neutrinos have approximately the same present-day density as the photons that make up the background radiation, and if they have a mass in the range of one ten-thousandth to one hundred-thousandth the mass of the electron, they could account for enough mass to close the universe. (The estimate was later confirmed by more detailed calculations.)

This point became particularly relevant in 1980 when V. A. Lubimov and his collaborators at the Institute of Theoretical and Experimental Physics in Moscow announced they had found evidence that the electron neutrino has a mass within that range. On the basis of this result it seemed neutrinos were ideal candidates to be the dominant mass in the universe. Since then, however, the likelihood that light neutrinos are the dark matter has become much smaller. In the first place, there are many outstanding experimental questions about the Soviet result; as a matter of fact, a recent finding by a group at the Swiss Institute for Nuclear Research appears to contradict it. In addition a great deal of work by astrophysicists has shown that theoretical pictures of a universe dominated by light neutrinos are not as compatible with observation as it once seemed.

The first such theoretical evidence came in 1979 from investigations by Scott D. Tremaine and James E. Gunn, then both at Caltech. They noted that, for reasons based partly on the Pauli exclusion principle, neutrinos in the relevant mass range could not condense sufficiently to be dark matter on scales much smaller than galaxies. The existence of dark matter on such scales has since been demonstrated convincingly by observations of dwarf spheroidal galaxies.

This work does not preclude the possibility that neutrinos are the dark matter on larger scales. Nev-

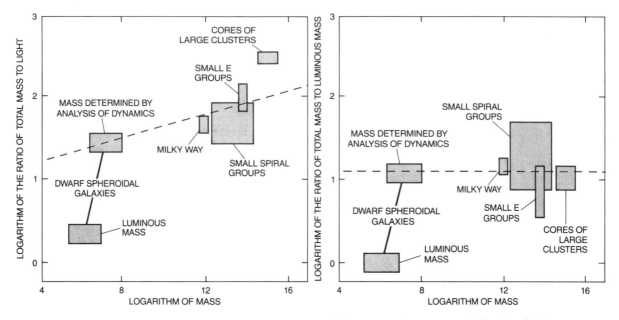

Figure 5 STRUCTURES ON DIFFERENT SIZE SCALES have different mass-to-light ratios (*left graph*), but about the same ratio of total mass to luminous mass (*right graph*), which seems to be constant, indicating that larger structures do not have proportionally more mass than smaller ones.

ertheless, such a proposal seems incompatible with substantial recent theoretical work describing the evolution of the early universe, which has demonstrated that the large-scale gravitational clustering (the clustering of galaxies and of clusters of galaxies) likely to occur in a neutrino-dominated universe does not seem to resemble the clustering actually observed.

In a neutrino-dominated universe the first structures to form would not be on the size scale of galaxies but rather on the scale of clusters of galaxies or even superclusters (clusters of clusters of galaxies). Unlike normal matter, neutrinos in the early universe were not coupled to electromagnetic radiation. Even so, for some time they were not able to clump together appreciably because, being extremely light they moved at relativistic speeds, and relativistic objects are not bound gravitationally except by very highly condensed objects such as black holes.

As the universe expanded, neutrinos cooled until they slowed down and became nonrelativistic. At the same time, the radiation background continued to cool to mean energies below those of the nonrelativistic neutrinos. Shortly before the time at which normal matter decoupled from electromagnetic radiation, neutrinos having masses in the appropriate range to close the universe would have become nonrelativistic and would have begun to make up the primary component of the energy density of the universe. Analytic calculations show that only after this time could they have clumped together gravitationally. At any earlier times, fluctuations on scales smaller than the horizon would have been broken up because the neutrinos, being relativistic, would not have been bound to dense regions.

Thus the first scale on which fluctuations could have grown in a neutrino-dominated universe is the scale of the horizon distance at the time when neutrinos could begin clumping gravitationally. This distance scale corresponds to the size of superclusters, not that of galaxies. Soon after it had decoupled, normal matter would have been drawn into the gravitational potential wells caused by clumps of neutrinos. These supercluster-size formations might then have fragmented into galaxies.

That scenario of a neutrino-dominated universe is attractive in many ways. It would have led to a system of filament-shaped superclusters and large "voids" (regions empty of matter) that resemble features identified in current surveys of large-scale clustering [see "Very Large Structures in the Universe," by Jack O. Burns: SCIENTIFIC AMERICAN, July, 1986]. In addition, the fact that gravitationally bound formations of neutrinos could begin to grow earlier than systems composed of normal matter indicates that the initial density fluctuations in the universe could have been small enough to be at least marginally consistent with measurements of the background radiation's isotropy.

These attractive features led Carlos S. Frenk of the University of Cambridge, Simon D. M. White of the University of Arizona and Marc Davis of the University of California at Berkeley and, independently, Joan Centrella of Drexel University and Adrian L. Melott of the University of Chicago to develop numerical models investigating the details of gravitational clumping in a neutrino-dominated universe. The investigators encountered serious difficulties when they tried to re-create the clustering that has actually been observed. Essentially they found that in a neutrino-dominated universe the fragmenting of clusters into galaxies and the formation of galaxies would have to have occurred relatively recently (when the universe was at least half its present age) in order to match the currently observed level of clustering. This conclusion is hard to reconcile with the existence of such structures as quasars, which formed in much earlier eras.

In general, the major problem with neutrino-dominated cosmology is that in order for galaxies to have condensed by the present time, structures on much larger scales would have to be much less diffuse than the observed large-scale structures actually are, because structure on the scales of galaxies and superclusters would have formed contemporaneously. Well-defined large-scale clustering would also cause difficulties in matching the predicted random velocities of galaxies in clusters to the observed velocities. For these and other reasons a neutrino-dominated universe now seems implausible.

A way out of the problems with neutrino models seems clear: find models in which galaxies can form significantly earlier than larger structures do. This suggests the need for what has become known as cold dark matter: dark matter that was so cold (that is, moving so slowly) that it was nonrelativistic significantly earlier than neutrinos were and could therefore cluster gravitationally much earlier.

The time at which a class of particles becomes nonrelativistic is a key factor in determining the size of structures that can be formed by that class of particles. At times before the particles become nonrelativistic, structures on scales smaller than the horizon would break up. Hence, in order for galaxies to form before larger structures, cold dark matter would have to have been nonrelativistic by the time the horizon reached the scale size of galaxies.

Ever since the problems with neutrino-dominated theories became clear, a great deal of effort has gone into the analysis of cosmology dominated by cold dark matter, and almost all the results have been positive. Because density fluctuations can grow earlier, the initial fluctuations need not be as large and so any conflict with the observed isotropy of the background radiation is eliminated. Moreover, because cold dark matter could have clumped on smaller scales than neutrinos could have, it might account for the excess mass in such small structures as dwarf galaxies.

Detailed analytical and numerical investigations are most encouraging. For example, it has been shown that the presence of cold dark matter in the early universe could account in detail for the shape and structure of many types of galaxies. More generally, Frenk and George Efstathiou of Cambridge, along with Davis and White, have shown numerically that clustering on large scales in a universe dominated by cold dark matter can match well with most of the observed features of the actual clustering.

There is still at least one obstacle that apparently prevents complete agreement between theory and observation if the universe is exactly closed and dominated by cold dark matter: Where is the matter? Apparently it can cluster readily on galactic scales, but, as I have described, there is no evidence for a critical density on such scales. One solution to the problem is to assume that galaxies themselves are not good indicators of where most of the high concentrations of mass are: that much of the cold dark matter lies in regions uncorrelated with the locations of these luminous systems. It could well be that galaxies represent statistically rare events, and that most of the mass in the universe has not ever condensed to form galaxies. Examining the clustering of galaxies would then give a biased value for the actual mass density of the universe. The implications of this proposal have been studied in detail, and it appears to lead to scenarios that agree well with most aspects of the observed clustering (with some notable exceptions). Moreover, current work by Frenk and his collaborators suggests that scenarios in which galaxies are statistically rare might arise more naturally from gravitational clustering than had previously been supposed.

The cold-dark-matter hypothesis has forged a strong link between particle physics and cosmology. At a time when cosmologists were deciding some form of cold dark matter was necessary, high-energy physicists were independently proposing the possible existence of new, exotic particles within the framework of various unified theories. As it happens, several of the particles proposed to fill theoretical gaps in high-energy physics could also serve quite naturally as the cosmologists' cold dark matter. These particles have the disadvantage that they have not been observed; unlike neutrinos, they are at this point merely theoretical constructs. Nevertheless, they have the virtue that their existence was proposed independently from cosmology: they were suggested as solutions to quite different problems in particle theory, and yet each of them, for entirely different reasons, could act as cold dark matter.

Among the most attractive candidates on the market today are particles called axions. The existence of axions follows naturally from a theoretical approach developed to explain a special relation that links, in the theory of strong interactions between quarks, the two kinds of symmetry known as charge conjugation and parity.

An interaction is said to be symmetrical under charge conjugation if the interaction would "look" the same were every particle to be replaced by its antiparticle (which has the opposite charge). An interaction is symmetrical under parity if it would look the same when mirror-reflected. The interactions governed by the strong nuclear force (the force that bind quarks together to form protons and neutrons) appear to be symmetrical to a very high degree under a special combination of charge conjugation and parity: the interactions look much the same if all the particles are replaced with their antiparticles and the entire interaction is mirror-reflected. Theoretically this special combination of symmetries need not hold true. The equations governing the strong interactions include several terms that could in principle grossly violate the combination of symmetries.

In 1977 Roberto D. Peccei and Helen R. Quinn, then both at Stanford University, suggested a way to explain why the combination of symmetries is obeyed so well. Their solution was to introduce a new kind of symmetry—a relation between the forms of different fundamental forces that is manifest at sufficiently high energies but is broken at low energies. It was later pointed out by Frank Wilczek of the University of California at Santa Barbara and Steven Weinberg of the University of Texas at Austin that the fact that the Peccei-Quinn symmetry breaks indicates the existence of a new, very light particle. The new particle is the axion. Much recent theoretical work has refined the original model and increased the temperature at which the Peccei-Quinn symmetry is expected to be broken. One of the big surprises to result is that, because the existence of axions depends on symmetry breaking, an axion "background field" might form in the universe, much as a background electric field would exist if the universe were not charge-symmetric (that is, if it did not contain equal numbers of positive and negative charges). Although axions are themselves very light, calculations show that the background field as a whole could clump in much the same way as heavier, nonrelativistic particles would, making the background field an ideal candidate for dark matter.

Another candidate for cold dark matter comes from the theoretical framework known as supersymmetry. In the theory of supersymmetry, for every particle now known there exists a "supersymmetric partner": a particle identical in most respects except spin. Such particles have not yet been observed in the laboratory, and so they must have large masses. Simple models suggest that supersymmetric partners could behave, in their interactions with normal matter, much like very heavy neutrinos. The most promising dark-matter candidate of the supersymmetric partners is the supersymmetric partner of the photon, which is called the photino. Calculations done by me and by others have shown that photinos in the mass range of from one to 50 times the mass of the proton could naturally have sufficient cosmic abundance to close the universe today. Although this proposal has generated a great deal of excitement recently, I should note that the models predicting the existence of photinos lead to other cosmological predictions that are hard to reconcile with observations.

A final candidate, related to the hit parade of cold-dark-matter candidates, is not a particle at all. It is a structure called a cosmic string. Cosmic strings are extended topological defects that might have arisen from symmetry breaking in the early universe. They would take the form of long, thin tubes of constant and very great energy density winding through the universe. Much work has gone into showing that cosmic strings could have evolved in such a way that their total energy density would be less than that required to close the universe. Nevertheless, in a universe dominated by cold dark matter and containing strings, the mechanism of galaxy formation, although it is quite different from mechanisms in standard cold-dark-matter models, might still lead to clustering that matched observations.

What makes all these dark-matter candidates so intriguing at present is the prospect that each of them may well be detected, directly or indirectly, in the near future. Experiments are possible that would rule out or, what is more significant, confirm various ones of the hypotheses. A positive result in any of these experiments would yield invaluable information about the evolution of large-scale structure in the universe and about the fundamental structure of matter, and it might provide a unique mechanism for probing the sequence of events that occurred during the first few seconds of the big-bang explosion itself.

Pierre Sikivie of the University of Florida was the first to point out that cosmic axions, although they interact with other matter extremely weakly, might be detected in microwave cavities (cavities in which electromagnetic radiation in microwave frequencies resonates). A background field of axions oscillating together might produce electromagnetic radiation that could in principle be detected in a microwave device. Wilczek, John Moody of the University of California at Santa Barbara, Donald E. Morris of the Lawrence Berkeley Laboratory and I have investigated this detection scheme in detail and have proposed refinements and alternative schemes. The sensitivity necessary to detect cosmic axions appears to be near the limit of modern technology, although the technology itself is improving rapidly.

Heavy dark-matter candidates, such as photinos, might be detected in several ways. Recently I suggested, as several other workers did independently, that heavy dark-matter candidates in the galactic halo could be captured in the cores of the sun and the earth, where they would accumulate. There, as later calculations have shown, they could collide with their antiparticles (which could also be captured) in annihilation reactions that could produce light neutrinos. The light neutrinos might then escape from the sun's or the earth's core and be measured in large underground detectors. The degree to which such a flux of light neutrinos has not yet been observed puts limits on the masses and densities of heavy dark-matter candidates.

Recently it has been pointed out that heavy dark-matter particles might also be detected directly by devices that are sensitive to very small deposits of energy in very large volumes of material. A variety of new detectors of this type have recently been proposed. One device, put forward by Blas Cabrera of Stanford, Wilczek and me, is designed to measure a small increase in the temperature of a large sample of ultracold silicon or of another pure crystalline material. The increase in temperature would occur when sound waves, or phonons, produced by impinging dark-matter particles, scattered and randomized. Work by Cabrera, Barbara Neuhauser and Jeffrey C. Martoff at Stanford suggests that the sound waves themselves could perhaps be detected directly.

In one class of possible detectors (see Figure 6), when an impinging dark-matter particle scatters off the nucleus of a silicon atom (6.1), it causes a set of phonons to spread throughout the material (6.2). Phonons arriving at the silicon's surface will have a distinctive pattern (6.3), which will depend on the location and intensity of the original collision. One detector configuration might detect individual phonons in the pattern as they impinge on the surface of the crystal. To do so the silicon could be overlaid with strips of two layers of superconducting aluminum sandwiching a layer of aluminum oxide (6.3a). In superconductors electrons are bound together in pairs called Cooper pairs. An incoming phonon might break apart a Cooper pair, and if the layers are kept at different voltages, the freed electrons might "tunnel" from one layer to the other, forming an electrical current (6.3b). Or, investigators could measure the rise in the silicon's temperature after the initial energetic phonons had dissipated into a uniform background of random thermal vibrations (6.4). Then the detector could consist of a thin film of a material whose electrical resistance increases sharply with temperature (6.4a). A change in temperature of the sample as a whole (6.4b) could be determined from the change in resistance.

Even cosmic strings may soon be detectable, either by their direct gravitational effects on the light from distant quasars and the microwave back-

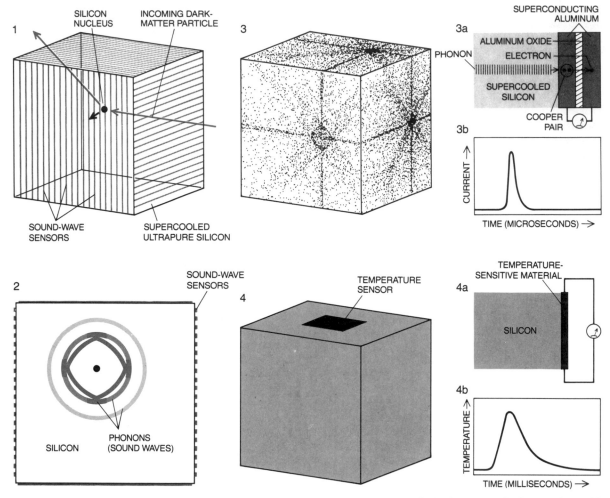

Figure 6 DETECTION DEVICES for dark-matter particles might be made of very pure silicon crystals cooled to within one degree of absolute zero. Such crystals could react measurably to extremely small deposits of energy.

ground (concentrations of energy as dense as cosmic strings should create gravitational fields that would bend light appreciably) or indirectly by measurement of the gravity waves or other radiation they should emit as they evolve.

The solution of the dark-matter question could have broad effects on many areas of physics and astronomy. At stake are fundamental notions about both cosmology and particle physics, and it is fitting that each field—often by provoking active debate in the other—has played an important role in the symbiotic evolution of this area of research.

It is important to recognize, however, that cosmology is in many ways in its infancy. There are comparatively few experimental and observational data available for theorists to work with, and so dramatic changes in the field are possible and much of the standard wisdom may be in error. The point is well illustrated by several new results that arose as this article was being written, any of which may have a profound effect on the field.

One new observational result is found in the preliminary analysis of a deep-sky survey being made by Margaret J. Geller, John P. Huchra and their collaborators at the Harvard-Smithsonian Astrophysical Observatory. It seems that nearby galaxies are clustered in filmlike surfaces that surround nearly spherical voids—a structure resembling that of soapsuds or foam bubbles. This remarkable observation, which could completely revise cosmologists' picture of large-scale structure, suggests that forces other than those of gravity are perhaps at work in determining the present-day large-scale structure.

In another new development, work done independently by Tremaine (now at the Canadian Institute of Theoretical Astrophysics) and J. Anthony Tyson at AT&T Bell Laboratories suggests that galactic rotation curves may not be flat indefinitely but rather may drop off at radiuses beyond about 30 kiloparsecs. The work implies that whatever makes up the dark matter may interact more strongly with normal matter than the cold dark matter would be expected to.

Finally, recent data on the motions with respect to the microwave background of very large-scale regions of matter have provided evidence that these regions are moving, together, with an extremely large drift velocity. No current theory of large-scale structure can explain this apparent phenomenon. New measurements such as these, as well as the possibility of detecting the dark matter itself, may soon revolutionize the accepted picture of the universe.

POSTSCRIPT

The dark-matter problem provides an explicit and timely demonstration of the "cosmic connection" between particle physics and cosmology that has blossomed in the past decade and that is the subject of this reprint volume. Many of the developments described in the subsequent articles here have impacted, in one way or another, our present perspective of the dark-matter issue.

Two factors form the basis of this intellectual connection. First, there is a distinct possibility that most or all of the dark matter in the universe is some exotic form of matter. This matter could be composed either of new types of elementary particles or made from coherent configurations of the fields associated with new forces between elementary particles. Beyond this, the distribution and abundance of dark matter today can reflect the nature of elementary-particle interactions at high en-

ergy. One of the most pressing problems in cosmology is to explain the origin of the observed macrostructure of the universe. This explanation is inevitably linked to our understanding of how the initial conditions of the big-bang expansion developed—a question that depends ultimately on our understanding of microphysics.

Several developments, which have taken place in the short time since my article appeared, attest to both the speed with which the interface between cosmology and particle physics is developing and the potential for radical changes.

Eugene Loh and Earl Spiller at Princeton apparently provided the first empirical support for the flatness of the universe. They claim to have measured the distances of about a thousand galaxies within a radius of about six billion light years—a fair fraction of the observed universe. Counting the number of galaxies at a given distance is one way to probe the curvature of space, because volume changes with distance in a way that depends on curvature. Their data appear to be consistent with a flat universe—allowing the possibility of three to five times greater mean density of matter than implied by all previous work. Since its appearance, the particular method of analysis they used has been seriously questioned by other workers on both experimental and theoretical grounds. While it seemed to signal the beginnings of agreement between theory and observation, it now appears as if this method cannot give definitive results.

Next, Davis and his collaborators are continuing their numerical simulations of clustering in a cold-dark-matter-dominated universe. The most recent results are very encouraging. Structures resembling, on a fine scale, the observed galactic mass distributions can form in a statistical ensemble that, on large scales, could reconcile a flat universe with observation. Moreover, a structure recently observed by Geller and collaborators bears a resemblance to certain (but not all) features of the fascinating bubble-like structure on large scales. At the same time, Alan Dressler and his colleagues have continued to argue that a large region of galaxies is falling coherently in the direction of the Hydra-Centaurus cluster. The large mass agglomeration required to produce this is not easily explained by any cold-dark-matter model.

The search also continues for direct and indirect signatures of cold-dark-matter candidates. New calculations of stellar production of light axions in the sun, red giants, white dwarfs, neutron stars and the recent supernova in the Large Magellanic Cloud have tightened the bounds on these candidates. Also, more refined calculations of the signal in the atmospheric neutrino background for the annihilation products of dark-matter candidates captured in the sun, earth and galactic halo continue to be produced. Proton-decay detectors, discussed in Chapter 7, "The Search for Proton Decay," by J. M. LoSecco, Fredrick Reines and Daniel Sinclair, have not yet seen such a signal, thereby strongly constraining a variety of heavy weakly interacting particles (WIMPs) as cold-dark-matter candidates. These constraints also seem to be supported by recent analyses from ultra-low background experiments on double beta decay of nuclei, which are sensitive directly to the scattering of very heavy dark-matter candidates. They have as yet yielded no such signal. As a result stable neutrinos with masses in excess of 15 to 20 GeV may be ruled out as candidates for galactic halo dark matter. Finally, at present at least 10 different groups around the world are building ultracold detectors of various types to directly probe for WIMPs, which should be operational soon.

The suggestion, made by myself and independently by several other people, that WIMPs captured in the sun could give rise to the observed apparent shortfall of solar neutrinos detected on earth has led to a brief flurry of activity. Recent work by John Faulkner and collaborators has suggested that this could result in a number of other observed solar features. Unfortunately, as I and my collaborators demonstrated, when analyzed in detail the capture of WIMPs from the galactic halo yields abundances that are generically too small for this mechanism to involve standard dark-matter candidates. Thus, even more exotic objects would be required.

There has also been a great deal of activity recently on the formation and evolution of large-scale structure in a cosmic-string-inhabited universe. This has focused on the ability of large cosmic strings to evolve via the formation of loops, which can subsequently decay quickly enough to yield acceptable mass densities today. Inside cosmic strings fundamental forces can appear in the same symmetric configuration they could have had at the high temperatures in the early universe. The different possibilities result in different possible observable consequences today. These scenarios could in principle explain galaxy formation and also yield general features of observed galaxy clustering.

Finally, several groups recently followed up on a possible loophole in the constraints from nucleosynthesis on the abundance of baryonic matter today. They suggest that under certain model dependent conditions, a flat universe dominated by normal matter would be consistent with the observed abundances of light elements resulting from primordial nucleosynthesis. If this were true all the dark matter could in principle be baryonic. Alternatively, another group has suggested that the decay of some heavy particle well after the standard nucleosynthesis epoch might alter primordial abundances in a way that would also allow the dark matter to be baryonic. Both these scenarios appear to suffer problems predicting the correct primordial abundance of the isotopes of lithium. However, the matter is still unresolved.

SECTION

II

THE UNIVERSE

. . .

The Structure of the Early Universe

The large-scale structure of the universe today is regular to within one part in 1,000. There is evidence that it has been that way since 10^{-35} second after the start of the big bang.

. . .

John D. Barrow and Joseph Silk

The cosmological principle is the powerful concept that the universe is homogeneous and isotropic. In other words, the large-scale features of the universe would appear the same to an observer in any galaxy no matter in which direction he looked. Much observational and experimental work supports the cosmological principle, which is deeply rooted in physics and natural philosophy. How did the universe acquire its large-scale uniform structure? Either it has pretty much always been that way or it was highly irregular and chaotic right after the big bang and has evolved into its present form because of certain smoothing and heating mechanisms. According to the latter possibility, called chaotic cosmology, the smoothing and heating mechanisms would give rise to the current regular universe regardless of the extent of the initial irregularity. Therefore chaotic cosmological theories eliminate the vexing problem of having to know the initial conditions of the universe.

Attractive as the elimination of the problem is, such theories may have a fatal drawback. We believe the proposed smoothing and heating mechanisms would irreversibly generate more thermal energy than seems to exist in the universe today. We think the universe as a result had only an infinitesi-

mal degree of irregularity at the time of its creation. What happened at the precise moment of creation is not yet known because unfamiliar physical principles unique to the immense densities and temperatures of that moment mask the initial structure of the universe. Matter behaves in a way that gives little inkling of what the universe was like in the first 10^{-35} second of its existence.

The cosmological principle gains observational support from the fact that the universe is undergoing an expansion in which every galaxy cluster is rushing away from every other. In 1923 Edwin P. Hubble discovered that the rate of expansion increases with distance from the observer. He detected the recessional motion of the distant galaxies through measurements of their optical spectra. The wavelength at which electromagnetic radiation from a distant object reaches the earth is increased by the velocity of recession of the object with respect to the observer. This is the well-known red shift, so named because if the radiation from the receding object is in the visible region of the spectrum, it is made redder.

The amount of the red shift is given as a number corresponding to the fractional increase in the wavelength of the received radiation. For example,

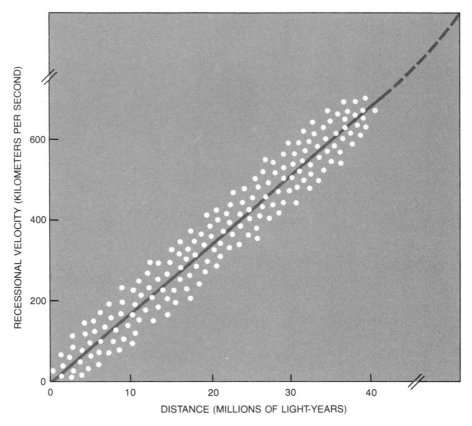

Figure 7 HUBBLE CONSTANT (*solid colored line*), or the increase in recessional velocity with distance, is about 17 kilometers per second per million light-years. The ratio of velocity to distance may depart from the constant for a galaxy (*white dots*) that has an orbital velocity because it is a member of a cluster. Because the recessional velocities have probably declined since the big bang, the velocity-distance ratio may have been larger at extreme distances (*broken colored curve*).

galaxies 20 million light-years away (among the closest to our own galaxy) have a red shift of .001 and galaxies 10 billion light-years away (among the most distant) have a red shift of .75. By measuring the red shift Hubble was able to calculate the recessional velocities of distant galaxies and thus the expansion rate of the universe.

The nature of the expansion can be understood by a traditional visual metaphor: likening the universe to a spherical balloon with dots painted on its surface, each dot representing a galaxy (see Figure 8). As the balloon is inflated the distance between any two dots (as measured on the surface of the balloon) increases at a rate proportional to the distance between them. An observer at any dot would see all the other dots receding from him

uniformly in all directions; no observer would occupy a privileged position. To put it another way, the expansion has no center.

It is not known whether the expansion of the universe will continue forever or whether someday the galaxies will stop receding from one another, start moving in the opposite direction and eventually fall together. Either possibility is consistent with prevailing cosmological theory, which maintains that the universe began with an explosion from a super-dense state. The type of expansion is determined by space-time geometry of the universe. Infinite expansion means an "open" universe; finite expansion followed by collapse means a "closed" universe. The critical intermediate case, advocated by Albert Einstein and Willem de Sitter in 1932, is where the universe has the minimum energy

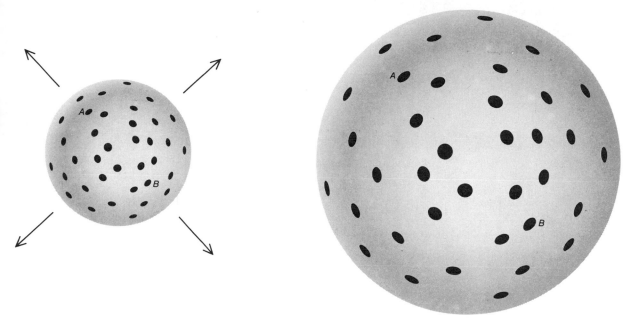

Figure 8 BALLOON MODEL OF THE UNIVERSE with each dot corresponding to a galaxy. When the balloon is inflated the distance between any two dots on its surface increases at a rate proportional to the distance, and the geometrical distribution of the dots does not change.

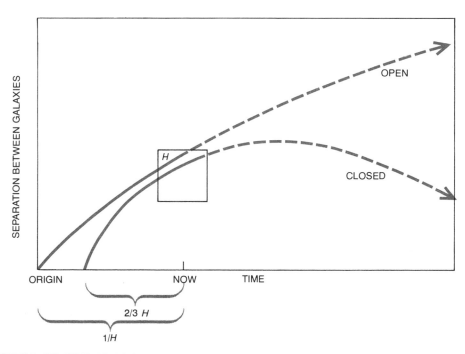

Figure 9 MODELS OF COSMIC EVOLUTION indicate how the separation between the galaxies changes with time. If the universe continues to expand forever, the age of the universe is the Hubble time: the inverse of the Hubble constant (H). If the universe eventually stops expanding and starts contracting, the age of the universe is less than $2/3\ H$.

needed to overcome the decelerating influence of gravity and expand forever to infinity. Whether the universe is open or closed is difficult to determine because the expansion energy is near the critical value. This is the subject of continuing investigation in astronomy.

The universe is isotropic and homogeneous not only in its rate of expansion but also in its distribution of constituent objects. Hubble counted the number of distant galaxies in separate quadrants of the sky and in different volumes of space. He found that the larger the volume is, the more galaxies it contains. Moreover, the distribution of galaxies with direction varied hardly at all. More recent surveys have probed the uniformity of the distribution of galaxies in the universe to much greater distances. For example, when distant regions with a radius of a gigaparsec, or 3×10^9 light-years, are compared, their populations of radio-emitting objects (galaxies and quasars) are found to be equal to within 1 percent.

The most compelling evidence of isotropy comes from the background of microwaves, or radio waves with wavelengths in the millimeter range, that seem to flood the entire universe. The radiation was discovered in 1965 by Arno A. Penzias and Robert W. Wilson of Bell Laboratories. Sky maps of microwave radiation based on measurements made by an-

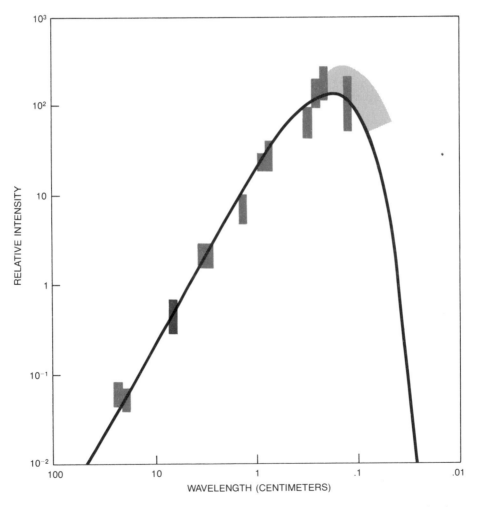

Figure 10 MEASUREMENTS OF RADIATION emitted by the big bang at microwave frequencies by Penzias and Wilson (*colored region*) and others (*shaded regions*) show that the spectrum of the radiation is the spectrum of a black body at 2.9 degrees K.

tennas carried to high altitudes in aircraft and balloons show that the intensity of the radiation is isotropic to better than one part in 1,000.

The homogeneity of the universe is more difficult to express quantitatively. The universe is of course highly inhomogeneous on a small scale such as that of the solar system or our galaxy. On a larger scale the homogeneity of its content of matter is indicated by the uniformity of the distribution of visible galaxies and radio sources and also by the uniformity of a universal background of X rays. On the scale of the entire observable universe the most sensitive indicator of homogeneity is the universal background of microwaves, which is homogeneous to better than one part in 1,000. The measurement is not definitive, however, because the radiation might have been rescattered by the intergalactic medium on its journey to the solar system and hence its variations might have been smoothed.

The microwave background radiation establishes much more than the validity of the cosmological principle for the recent history of the universe. The spectrum of the radiation is identical with the one that would be generated by a black body (a perfect emitter of radiation) at a temperature of 2.9 degrees Kelvin (2.9 degrees Celsius above absolute zero). The radiation is today only a feeble glimmer, but it attests to a fiery past. For the radiation to have the spectrum of a black body the early universe must have passed through a hot, dense phase. One of the most remarkable predictions of modern cosmology was the suggestion made by George Gamow (and his co-workers Ralph A. Alpher and Robert Herman) that a remnant of the big bang should still be visible as a pervasive background of black-body radiation. Gamow believed all the chemical elements with the exception of hydrogen could have been created in the hot, dense phase of the universe just after the big bang. To him the universe was a giant fusion reactor, and the simple requirement that the universe not immediately burn all its hydrogen into helium led directly to the prediction that the radiation, although greatly cooled and diluted by the expansion, should still be present with a temperature of about five degrees K.

Today it is known that the universe did not stay hot and dense long enough for the heavy elements such as carbon and iron to be built up in successive reactions from primordial protons and neutrons. It is now known that the heavy elements were synthesized in the interior of stars. The oldest stars seem to be deficient in heavy elements, which means that at the time of their formation their environment was poor in such elements. Yet the helium abundance of the old stars is essentially the same as that of much younger stars rich in heavy elements. Moreover, many kinds of galaxies are alike in their helium abundance. The uniform universal distribution of helium, which is second only to hydrogen in its cosmic abundance, indicates that it was chiefly created not in stellar interiors but in the hot aftermath of the big bang, as Gamow visualized. The general agreement between his prediction of a black-body radiation of five degrees K. and Penzias and Wilson's discovery of radiation of 2.9 degrees K. and the accurate prediction of the primordial helium abundance are the most compelling arguments for a hot big bang.

From data on the microwave background radiation theorists were able to calculate a new fundamental quantity: the ratio of the number of photons (the massless particles of electromagnetic radiation) in the universe to the number of nucleons (the massy protons and neutrons). The ratio is about 10^8 photons to one nucleon and is a measure of the average thermal entropy associated with each nucleon. Entropy, usually represented by the letter S, is defined as a number that indicates how many states are possible in a system. To put it another way, entropy is a measure of randomness or disorder. A liquid is an example of a high-entropy system because its atoms can arrange themselves in a huge variety of ways; a crystal lattice is an example of a lower-entropy system because its atoms are arranged in a highly ordered way. The ratio of the density of photons to the density of nucleons averaged over a large volume of the universe is a measure of the average entropy because photons constitute the most disordered states of thermal energy and nucleons constitute the most ordered states. Hence the relative abundances of these two kinds of extreme state are a measure of the average entropy.

According to the second law of thermodynamics the total entropy of the universe increases continuously as time goes on. This means that at the time of the big bang S was less than 10^8. The isotropic expansion of the universe did not dissipate much heat, so that any increase in S must be due to other mechanisms. An entropy of 10^8 is quite large compared with the entropy of about 1 exhibited by systems in the terrestrial environment. This means that the universe as a whole is a relatively hot place. A physicist's first reaction to a hot system with a

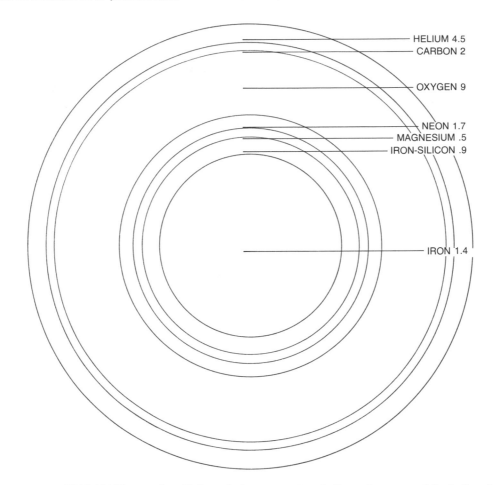

HELIUM 4.5
CARBON 2
OXYGEN 9
NEON 1.7
MAGNESIUM .5
IRON-SILICON .9
IRON 1.4

Figure 11 A STAR WITH A MASS more than 20 times that of the sun is unstable and explodes as a supernova. As it contracts under the mutual gravitational attraction of its constituents the pressure and temperature of its interior increase until thermonuclear reactions first transmute hy-drogen into helium, then some of the helium into carbon. The temperatures continue to rise until carbon starts to be transmuted into heavier elements. At each stage a shell of lighter material is left behind. The masses of the shell are in solar masses.

high degree of regularity is to suppose much thermal energy was dissipated in the history of the system, the heat and the regularity being the aftermath of frictional smoothing processes. We shall try to demonstrate, however, that for the universe this interpretation is unlikely to be correct.

By extrapolating backward from the present-day expanding universe to the time before galaxies formed, cosmologists have traced the origin of the universe to a singularity: a state of apparently infinite density. The singularity represents the origin of space and time perhaps 10 billion years ago. Before that time the laws of physics known today did not apply. Does that testify to the physicist's knowledge or to his ignorance? Before 1965 cosmologists debated the physical significance of this singularity. Some theorists thought it might be a spurious manifestation of the particular coordinate system chosen to describe the dynamics of the expansion of the universe, a manifestation similar to the singularity in geography. On a terrestrial globe there is a coordinate singularity at the North and South poles, where the grid squares of longitude and latitude vanish as the meridians of the globe intersect. Yet the fabric of the world does not physically break down at the poles.

Since 1965 several theorists have independently

shown that the cosmological singularity is not the result of a poorly chosen coordinate system. On the contrary, the singularity seems to be general, physically real and an inevitable consequence of the fact that gravity is attractive and acts indiscriminately on everything, including photons. The singularity was probably characterized by infinite density and curvature, although all that is known with certainty is that a material observer moving back into the past would experience an abrupt and disconcerting end to his trip through space-time when he encountered it. He would be unable to travel any farther because the laws of physics require the universe to have a space-time boundary. (The traveler need not, however, experience the big bang with its infinite densities and temperatures.)

The cosmological singularity is similar to the singularity in the event horizon of a black hole, the hypothetical surface in which matter and light rays are confined by gravity. Nothing leaves a black hole because the velocity needed to escape the grasp of its gravity is greater than the speed of light, which the laws of physics require cannot be exceeded. It has been shown that an unfortunate astronaut who fell freely into a black hole would reach a physical singularity within a finite time (as time would be measured on his own watch). The singularity would be invisible to external observers since it lies inside the event horizon of the black hole. Now, since photons feel the pull of gravity, it is possible to determine whether at some time early in the history of the universe the photons that currently make up the microwave background radiation could have exerted sufficient gravitational pull to create a trapped region analogous to a black hole. Within that region would be a cosmological singularity, which the physicist defines as the beginning of the universe.

That the universal expansion originated at a singularity has a far-reaching consequence in terms of the points in the universe that causally influence one another. Since no signal can move faster than the speed of light, an observer can be affected only by events from which a photon would have time to reach him since the beginning of the universe. Such events are described as lying within the observer's horizon (see Chapter 3, Figure 18). Consider two points with a spatial separation x at the beginning of time. Before the time it takes for light to travel between them (x/c, where c is the speed of light) the points will not "see" each other, "know" of each other's existence or be able to affect each other in any way. In general, regions of the universe with a spatial separation greater than ct will not know of each other's existence until a time t has elapsed. What this means is that regions of the isotropic microwave background in different directions of the sky (say more than 30 degrees apart) could never have causally influenced each other at any time in the past. That creates a paradox: How did causally disjoint primordial regions of the universe come to have the same temperature and expansion rates today to within at least one part in 1,000?

A further twist to the conundrum of the origin of the uniform background radiation comes from the fact that although the universe is quite regular on the scale of several tens of millions of light-years, it does have on a smaller scale some spectacular inhomogeneities in the form of galaxies and clusters of galaxies. The strength of the gravitational field exerted by the largest of these inhomogeneities suggests that their ancient precursors would have created an anisotropy in the microwave background radiation over a scale of a few angular degrees. Observational astronomers are currently searching for such an anisotropy. Anisotropy would also arise from the remnants of ancient directional disparities in the universal expansion rate. It is not clear what physical mechanisms have smoothed the irregularities into the structured universe of today.

What cosmologists are trying to account for are the entropy and the large-scale structure of the universe, which appear to have existed when the universe was less than a minute old. Such epochs are best defined, however, not in temporal units such as minutes or years, which are subject to correction as the yardsticks of astronomy are refined, but in units of red shift, which express the amount by which the universe was compressed with respect to its present size.

On the one hand there are "chaotic" cosmologists, who like biologists maintain that the properties of the universe are the result of evolutionary processes. They have tried to show that a kind of gravitational natural selection could deliver the present large-scale structure as the inevitable result of physical smoothing and heating processes that have been going on since the big bang. If it could be demonstrated that the present structure would have arisen no matter what the initial conditions were, then the uniqueness of the universe would be established in theory as well as in actuality.

On the other hand there are "quiescent" cosmologists, who appeal largely to the initial conditions to explain the present structure of the universe.

They hypothesize that when the universe was created at the singularity, it had certain definite and preferred structural features for reasons, say, of self-consistency, stability or uniqueness. This means that gravitational evolutionary processes played a role not in shaping the overall configuration of the universe but only in molding substructures such as galaxies, stars and planets. A good deal of the theoretical work in cosmology over the past decade has centered on finding ways of distinguishing between the two alternative cosmologies. We shall devote the remainder of our discussion to this work.

It is now believed that at ordinary temperatures all natural phenomena are governed by four fundamental forces: the gravitational force, the electromagnetic force, the weak force and the strong (or nuclear) force. These forces in conjunction with a small number of additional parameters such as particle mass determine the structural characteristics of the universe. As the history of the universe is traced backward in time from the present to the singularity at a red shift of infinity some 10 to 20 billion years ago, each of the four fundamental forces will at some point come to dominate the others. Today the

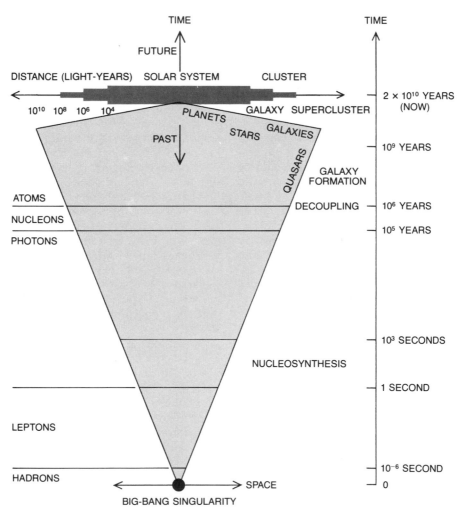

Figure 12 SPACE-TIME DIAGRAM of the universe traces it history from the singularity of about 20 billion years ago. The shaded region represents the horizon of a hypothetical observer at the singularity. Each fundamental force came to dominate the behavior of matter at some point.

gravitational force is the one that governs the dynamics of large-scale expansion. Although the gravitational attraction of two protons is 10^{40} times weaker than their electromagnetic repulsion, gravity becomes increasingly important in a system with a huge number of particles such as the universe. In the universe the number of positively charged particles should be equal to the number of negatively charged particles, and so on the whole the attractive and repulsive electromagnetic interactions cancel out and exert no significant long-range forces over large regions. Since gravity is only attractive, however, it plays a major role in massive systems. It is the influence of the gravitational force, which has infinite range and acts on the photons of radiation as well as on the particles of matter, that determines the size of the largest objects in the universe, such as planets, stars, galaxies and clusters of galaxies.

If a local region of the early universe happened to have a density higher than that of the surrounding regions, it would gravitationally attract more matter than the less dense regions. As it contracted under the influence of its own gravity, it would increase in density and attract matter even more efficiently. What began as a fluctuation in a fairly homogeneous universe would eventually snowball into a huge inhomogeneity. Galaxies seem to have formed at times equivalent to red shifts between 10 and 100 and to have come together in clusters at later times. It is not known with any certainty, however, whether galaxies formed out of fragments of much larger fluctuations that disintegrated or out of smaller fluctuations that came together because of their mutual gravitational attraction.

At a time equivalent to a red shift of 1,000 (about 300,000 years after the big bang) gas pressure is stronger than gravity over a dimension equivalent to that of about 100,000 suns, but gravity is much stronger than gas pressure over larger dimensions. Fluctuations on the order of these larger dimensions grow until they eventually become large enough to collapse and form bound objects spanning a range of masses from those of globular star clusters to those of galaxies. At times equivalent to red shifts greater than 1,000 the chief source of pressure is not gas pressure but the pressure of thermal radiation. In that epoch the dynamical behavior of density perturbations is determined by the electromagnetic force. Under these circumstances photons form a viscous fluid that inhibits the movement of electrons and protons. An electron, for example, would scatter impinging photons, which it feels as electrical pulses. Because of the law of the conservation of momentum the scattering would slightly alter the electron's trajectory. The net result is that the electron, locked into the radiation field by the ceaseless barrage of photons, cannot go anywhere. Once the electron joins a proton to form a hydrogen atom, however, it effectively no longer feels photons because it is interacting primarily with the electric field of the proton. The lack of movement of individual electrons means that fluctuations in the density of matter, called isothermal density fluctuations, are preserved until the time when electrons and protons combine to form electrically neutral atoms. Such atoms can travel freely through the radiation, and so the gravitational growth and collapse can proceed. The isothermal density fluctuations start to collapse when they acquire a mass of more than 100,000 suns. Such fluctuations come together to form galaxies.

Objects could also form from another kind of perturbation, called an adiabatic fluctuation, in which matter and radiation are perturbed together. If matter and radiation were squeezed slightly in a confined space, the excess pressure would create a kind of sound wave. Yet just as sound waves in air eventually dissipate and fade, so would a primordial sound wave. The critical length and mass below which the wave would be damped completely is determined by the ability of photons to escape from the adiabatic fluctuation in the time since the beginning of the universe. By a time equivalent to a red shift of 1,000 only adiabatic fluctuations more massive than the observed size of a massive galaxy or a group of galaxies survive the damping. In other words, less massive adiabatic fluctuations are smoothed out, whereas the more massive ones perhaps survive, grow and eventually collapse into massive galaxies and groups of galaxies.

In summary, the reconstruction of the probable evolution of the hot early universe consists of two kinds of density fluctuation, the isothermal and the adiabatic, which roughly correspond respectively to a globular star cluster or a dwarf galaxy and to a cluster of galaxies. The picture is undoubtedly a gross oversimplification because in general an arbitrary inhomogeneity would be expected to have an admixture of both isothermal and adiabatic components. Moreover, newly formed structures could merge or fragment, leaving no trace of their previous individual identity. In spite of these qualifications cosmologists would be rash to ignore this simple picture because the preferred masses that emerge out of the two kinds of fluctuation are comparable to the masses of the objects whose origin they are trying to explain.

At a time equivalent to a red shift of 10^{10}, when the universe was only a few seconds old and its temperature was 10^{10} degrees K., physical processes are mediated by the weak force. This force governs certain radioactive decay processes involving a free neutron or a neutrino: a spinning pointlike particle with no charge and negligible mass. At times equivalent to red shifts greater than 10^{10} the weak force keeps the protons and neutrons in thermal equilibrium: a statistical state where the number of particles with properties (position, mass, energy, velocity, spin and so on) in a specified range remains constant because the rate of particles entering that range is equal to the rate of particles leaving it. When particles achieve a state of thermal equilibrium, their behavior is determined not at all by their history but entirely by a set of statistical laws based on their temperature. This means that it is unnecessary to pry any farther into the past to understand the behavior and the relative concentrations of protons and neutrons.

The weak force in conjunction with the big-bang model of expansion leads to the prediction that the abundance of primordial helium in the universe is between 25 and 30 percent. This prediction, which has now been confirmed, has led to a precise and direct observational test of big-bang cosmology. What the success of the prediction means is that a few seconds after the big bang the universe was at least as regular on a large scale as it is today and had almost the same entropy. Cosmologists are trying to determine the degree of regularity before that time.

At times equivalent to red shifts greater than 10^{10}, neutrinos and their antiparticles (antineutrinos) play a big role. Today these particles are quite elusive because they almost never interact with anything in the rarefied medium of the present universe. Yet when the universe was a little less than a second old matter and radiation were so dense that neutrinos rapidly interacted strongly with them. At a time equivalent to a red shift of 10^{10} a typical neutrino traversed a significant extent of the universe before it collided with another particle. This means that the neutrinos could effectively transport energy and momentum over extremely large distances. They would do so by absorbing energy in high-energy regions of the universe and transferring it by occasional collisions to low-energy regions. As a result the neutrinos act to smooth out any nonuniformities in the distribution of matter that might have been created by directional differences in the overall expansion of the early universe.

The possibility of smoothing by neutrino-transport processes was suggested in 1967 by Charles W. Misner of the University of Maryland in the hope that such processes could remove a host of irregularities associated with the initial singularity of a chaotic universe. This hope has now foundered on the realization that if the initial expansion were sufficiently asymmetrical, the universe would expand so rapidly that there would not be enough time for neutrinos to collide with other particles. In other words, a highly anisotropic universe could remain that way. It seems that neutrino transport and other smoothing processes could only remove anistropies below a certain level. No matter how efficient a smoothing process was postulated, it was always possible to imagine a model universe that would nonetheless remain much more irregular than the present-day universe actually is. This consequence of the weak force cannot satisfy the central tenet of chaotic cosmology: the evolution of the regular present-day universe from any initial state no matter how irregular.

It is time to leave the epoch of the weak force to move closer to the cosmological singularity and tentatively probe the first milliseconds of the universe. There, where the temperatures and the particle energies exceed those achieved by the most powerful man-made accelerator and the radiation density is comparable to the density of the atomic nucleus, the dominant force between particles is the strong force. The extrapolation of our model into these earliest times is somewhat precarious because the understanding of the basic physics is not complete. When the strong force annihilates a proton and an antiproton, it gives rise to two energetic photons moving in opposite directions. In the first millisecond of the universe the temperature would have been so high that such an annihilation and the inverse process, the spontaneous production of nucleons and antinucleons from photons, would be quite efficient. Nucleons and radiation would have been indistinguishable. The average entropy of 10^8 photons per nucleon observed today implies that when the last annihilation took place, one proton survived for every 10^8 photons created by the destruction of other pairs of particles and antiparticles.

It seems that just before the universe was a millisecond old there was a minute imbalance between matter and antimatter: 1.00000001 particles per antiparticle. Until recently the origin of this peculiar imbalance was a complete mystery because of a principle having to do with baryons: heavy particles, including nucleons, that feel the strong force.

INTERACTION	SOURCE	FIELD QUANTUM	RELATIVE STRENGTH NOW	RANGE (CENTI-METERS)	MANI-FESTATION	UNIFICATION ENERGY (ELECTRON VOLTS)
ELECTRO-MAGNETIC	ELECTRIC CHARGE	PHOTON	10^{-12}	∞	ATOMIC AND MOLECULAR FORCES, ELECTRICITY	10^{11}
WEAK	LEPTONS MESONS	(W BOSONS)	10^{-14}	10^{-15}	RADIOACTIVE BETA DECAY	10^{23}
STRONG	BARYONS MESONS	PION, KAON	1	10^{-13}	NUCLEAR FORCES	10^{28}
GRAVITA-TIONAL	MASS-ENERGY	(GRAVITON)	10^{-40}	∞	LARGE-SCALE DYNAMICS OF MATTER	

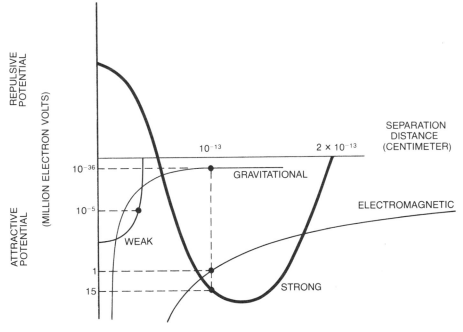

Figure 13 CHARACTERISTICS OF THE FOUR FORCES that govern all the interactions of matter and energy are given in the table; the range and relative strengths of the forces are given in the diagram. Above the unification energy the forces lose their individuality. Subatomic particles all respond to at least one of the forces. The particles fall in two broad categories: hadrons, which seem to have an inner structure, and leptons, which have no constituent parts.

Physicists believed the number of baryons in a system minus the number of antibaryons was absolutely fixed for all time. No interactions or transformations of the particles could ever change this quantity. If this were true for the universe as a whole, the asymmetry of one part in 10^8 of matter over antimatter must have been built into the initial structure of the big bang.

In recent times cosmologists have been actively investigating the consequences of a new extension of the theory of matter in which the electromagnetic, the weak and the strong force are all unified at sufficiently high temperatures. Above 10^{26} degrees K. these forces lose their individuality, whereas at lower temperatures they seem to be independent (although they are actually different aspects of an underlying unity). This kind of unification is possible only if the quarks that make up protons and other elementary particles are able to decay. Such decay has a surprising consequence: the proton is an unstable particle, although it has an average lifetime of about 10^{31} years. In spite of the infrequency of proton decays, physicists hope to observe some decay eventually with an experiment examining a sufficiently large mass: 1,000 tons of matter consisting of roughly 10^{32} protons.

The possibility of such unusual processes indicates that the level of particle-antiparticle asymmetry in the universe, which determines the observed entropy, is not absolutely invariant. It can change dramatically in the first 10^{-35} second of the universe, when the processes that mediate the decay of protons are abundant. Recent work shows that after this early instant a stable level of asymmetry between particles and antiparticles is eventually frozen into the universe, and the predicted value is close to the observed asymmetry of one part in 10^8. The observed entropy need not be strongly dependent on the initial conditions of the big bang in the first 10^{-35} second.

So far we have been taking into account only the corpuscular properties of matter. As we speculate on what happened right after the big bang the wave properties of matter must also be considered. According to quantum mechanics, every particle behaves as a wave with a length equal to 2.1×10^{-37} divided by the particle's mass. This wavelength, called the Compton wavelength, is infinitesimal by everyday standards, but only 10^{-23} second (the Compton time) after the big bang the Compton wavelength (10^{-13} centimeter) of a proton would be equal to the size of the causally connected region of the universe. There is nothing fundamental about this scale, however, because there is nothing fundamental about protons, which are of course made up of quarks.

The ultimate barrier is reached at times close to 10^{-43} second (the Planck time) after the big bang, because causally connected regions of the universe were compressed to a scale smaller than the Compton wavelength of their entire mass content. Before the Planck time the usual interpretation of spacetime is probably invalid because quantum-mechanical fluctuations dominate the geometry of spacetime. Undoubtedly many secrets of the universe would be revealed by an understanding of pre-Planck time, but achieving such an understanding is currently a remote possibility. Achieving it will probably have to await a new physical theory that synthesizes the theory of relativity and quantum theory. Cosmologists now regard the Planck time as being in effect the moment of creation of the universe; they leave to speculation any possibility of an earlier phase of evolution.

The epoch between the Planck time and the Compton time is a little more accessible to theoretical work. In that epoch quantum mechanics points to a mechanism that might have erased irregularities in the universe. According to quantum mechanics, all space is filled with pairs of "virtual" particles and antiparticles. Such particles, which materialize in pairs, separate, come back together and annihilate each other, are called virtual because unlike real particles they cannot be observed directly by particle detectors, although their indirect effects can be measured. If a pair of virtual particles is subject to a force field that is either extremely powerful or rapidly varying, its components might separate so quickly that they could never come back together. In this case the virtual particles would become real ones, their mass being supplied by the energy of the force field. Close to the Planck time the required force field might be generated by the changing expansion dynamics of the universe itself.

The Russian astrophysicist Yakov B. Zel'dovich has proposed that the production of real particles from virtual ones would erase the anisotropies and nonuniformities in the initial structure of the universe. It is imagined that radiation and particles would be preferentially spawned in the overly energetic regions of space. The newly formed particles would transfer energy from high-density regions to lower-density ones. Moreover, they would tend to equalize the rates of expansion in different directions. Much as people stepping onto a rotating merry-go-round tend to slow it down, the sudden appearance of particles would slow down the rapidly spinning and moving regions. Perhaps it is this quantum-physical mechanism that is responsible for the homogeneous large-scale structure of the present-day universe.

In the above discussion we have reviewed several smoothing mechanisms that might be responsible for the present regularity of the universe, mechanisms such as neutrino transport and the creation of real particles from virtual ones. As promising as these mechanisms may seem, we believe thermodynamical considerations demonstrate that they are severely limited. Since entropy increases with time, the present level of entropy in the universe (10^8) puts an upper limit on the amount of dissipation that has occurred during the past history of the universe. The smoothing of anisotropies and inhomogeneities in the initial structure would irreversibly convert the energy of irregularity into heat energy. The erasure of primordial chaos would generate radiation, but the low level of the microwave background radiation means that there could not have been an arbitrarily large amount of heating and smoothing in the past.

Moreover, it has been shown that in general the sooner after the big bang the irregularities are damped out, the more thermal radiation would be generated. Most of the dissipative mechanisms we have discussed would be fully operative right after the big bang. This means that if the universe were made with anything but a small degree of irregularity, entropy would be created at the singularity in excess of the observed level. But how small must the initial irregularity have been? The work of Barry Collins and Stephen Hawking of the University of Cambridge shows that a highly but not perfectly regular universe is unstable. The slightest deviation from regularity would tend to grow in time as the universe expanded regardless of the dissipative mechanisms. To put it another way, a universe beginning its expansion in anything but a precisely regular configuration would tend to become increasingly irregular. This means that the initial irregularities could only have been infinitesimal. Our own work indicates that there is a hierarchy of irregularities in the universe. As progressively larger volumes of space are examined, the degree of irregularity decreases in a way that suggests that initial

irregularities were only statistical fluctuations of a regular state.

What we are saying is that the present entropy level shows that the universe has evolved in an exceedingly regular way from as far back as the first 10^{-35} second of its existence. Before that time complex processes involving the nonconservation of the symmetry of particles and antiparticles and the quantum properties of the gravitational field erased any memory of the initial entropy per nucleon. Although the dissipation of chaos at those early times could also have generated many photons, nucleons would have been created as well, and the net number of photons per nucleon might have decreased or increased. Of course, the total entropy of the matter and radiation in the universe must always increase.

An unsettled question is whether the initial conditions are unique or whether another set of conditions could have done the same job. Several candidates for initial conditions have been proposed by theorists. One possibility is that the strong force served to keep matter highly rigid at the time of the singularity. At extremely high densities two nucleons could repel each other in the same way that two like magnetic poles do. Such a repulsion, which would prevent heavy nuclei from collapsing, might come to dominate the overall behavior of the interacting particles. An early state dominated by the strong force would remain quite regular because the high pressure would prevent distortions or turbulence from developing as the density increased.

Another possibility, developed by Roger Penrose of the University of Oxford, is based on the proposal that the overall gravitational field of the universe itself has an entropy that is proportional to and dependent on its uniformity. Since gravitational entropy, like all other forms of entropy, should always increase with time, the initial state of the universe would have been one of low gravitational entropy and regularity. To minimize gravitational entropy the universe might have acquired a regular, isotropic configuration, as a soap bubble minimizes the entropy associated with its surface area by assuming an isotropic spherical shape. As the universe ages and expands, the gravitational entropy would increase to reflect the growth of such inhomogeneities as galaxies and clusters.

Still another speculative possibility for the initial conditions of the universe is based on Mach's principle: The motion of an object is determined not by

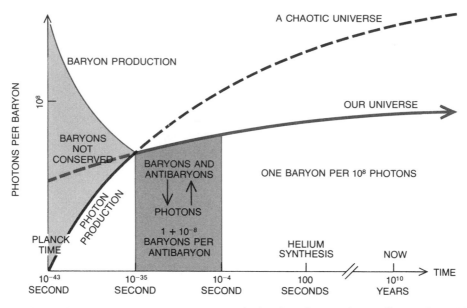

Figure 14 THERMAL EVOLUTION OF THE UNIVERSE in terms of the number of photons per baryon. Before 10^{-4} second there was an imbalance of matter and antimatter that may have arisen spontaneously at about 10^{-35} second. A chaotic universe after 10^{-35} second would have given rise to many more photons per baryon than are observed today.

the characteristics of some "absolute" geometrical space but solely by the material content of the universe. (The principle was first advanced in 1893 by Ernst Mach in a critique of Newton's concept of an absolute space to which the motion of all objects is referred.) Although Mach's principle is not a consequence of the theory of relativity, it has been introduced into the theory either as a boundary condition on the relativistic equations or as a sieve for eliminating solutions that are not physically acceptable. Derek Raine of the University of Leicester had developed a detailed version of the latter alternative, and out of his work comes the requirement that the infant universe be almost completely isotropic and homogeneous.

The last possibility we shall discuss is based on the idea that the initial conditions are limited by the very fact that they have led to human life on the earth. This idea, called the anthropic cosmological principle, was introduced by G. J. Whitrow of the University of London, Robert H. Dicke of Princeton University and Brandon Carter of the Meudon Observatory and was developed by John A. Wheeler of the University of Texas at Austin. Consider how this principle bears on the question of the size of the

universe. Since the universe is constantly expanding, its size depends on its age. The anthropic cosmological principle has convinced us that the universe must inevitably be about 10 billion light-years in diameter. A smaller universe would have existed for less than the billion years necessary for the heavy elements essential to human life to be synthesized by thermonuclear reactions in the interior of stars. Moreover, if the universe were much bigger and hence much older, the stars needed to establish the conditions of life would have long since completed their evolution and burned out.

The anthropic cosmological principle bears on the universe's entropy level of 10^8. If this number were increased by a factor of 1,000 or so, it would not be possible for protogalaxies to condense to the density at which stars would form. Without stars the solar system and the heavy elements of living matter would not have been created. If the universe were initially fairly irregular, it would have irreversibly generated copious quantities of heat radiation because of the many efficient entropy-generating channels open to it at the time of the singularity. Again this would have resulted in an entropy and a radiation pressure in excess of the values favoring the condensation of protogalaxies. Such a universe could not have been observed by us. These ideas indicate one thing: man's existence is a constraint on the kinds of universes he could observe. Many features of the universe that are remarkable to ponder are inevitable prerequisites of the existence of observers.

This unusual approach has even been extended to the values of the fundamental constants of nature. Consider a change in the strength of the strong-interaction coupling constant of only a few percent. As small an increase as 2 percent would block the formation of protons out of quarks and hence the formation of hydrogen atoms. A comparable decrease would make certain nuclei essential to life unstable. By the same token small changes in the electric charge of the electron would block any kind of chemistry and rule out the existence of stable planet-supporting stars.

Although the anthropic cosmological principle indicates why the structural features of the universe are in some sense inevitable, it leaves the reason for these features a mystery. Whatever the scientific status of the anthropic cosmological principle may be, its impact on the history of ideas may be significant. The principle overcomes the traditional barrier between the observer and the observed. It makes the observer an indispensable part of the macrophysical world.

COSMIC TIME	EPOCH	RED SHIFT	EVENT	YEARS AGO
0	SINGULARITY	INFINITE	BIG BANG	20×10^9
10^{-43} SECOND	PLANCK TIME	10^{32}	PARTICLE CREATION	20×10^9
10^{-6} SECOND	HADRONIC ERA	10^{13}	ANNIHILATION OF PROTON-ANTIPROTON PAIRS	20×10^9
1 SECOND	LEPTONIC ERA	10^{10}	ANNIHILATION OF ELECTRON-POSITRON PAIRS	20×10^9
1 MINUTE	RADIATION ERA	10^9	NUCLEOSYNTHESIS OF HELIUM AND DEUTERIUM	20×10^9
1 WEEK		10^7	RADIATION THERMALIZES PRIOR TO THIS EPOCH	20×10^9
10,000 YEARS	MATTER ERA	10^4	UNIVERSE BECOMES MATTER-DOMINATED	20×10^9
300,000 YEARS	DECOUPLING ERA	10^3	UNIVERSE BECOMES TRANSPARENT	19.9997×10^9
$1-2 \times 10^9$ YEARS		10–30	GALAXIES BEGIN TO FORM	$18-19 \times 10^9$
3×10^9 YEARS		5	GALAXIES BEGIN TO CLUSTER	17×10^9
4×10^9 YEARS			OUR PROTOGALAXY COLLAPSES	16×10^9
4.1×10^9 YEARS			FIRST STARS FORM	15.9×10^9
5×10^9 YEARS		3	QUASARS ARE BORN; POPULATION II STARS FORM	15×10^9
10×10^9 YEARS		1	POPULATION I STARS FORM	10×10^9
15.2×10^9 YEARS			OUR PARENT INTERSTELLAR CLOUD FORMS	4.8×10^9
15.3×10^9 YEARS			COLLAPSE OF PROTOSOLAR NEBULA	4.7×10^9
15.4×10^9 YEARS			PLANETS FORM; ROCK SOLIDIFIES	4.6×10^9
15.7×10^9 YEARS			INTENSE CRATERING OF PLANETS	4.3×10^9
16.1×10^9 YEARS	ARCHEOZOIC ERA		OLDEST TERRESTRIAL ROCKS FORM	3.9×10^9
17×10^9 YEARS			MICROSCOPIC LIFE FORMS	3×10^9
18×10^9 YEARS	PROTEROZOIC ERA		OXYGEN-RICH ATMOSPHERE DEVELOPS	2×10^9
19×10^9 YEARS			MACROSCOPIC LIFE FORMS	1×10^9
19.4×10^9 YEARS	PALEOZOIC ERA		EARLIEST FOSSIL RECORD	600×10^6
19.55×10^9 YEARS			FIRST FISHES	450×10^6
19.6×10^9 YEARS			EARLY LAND PLANTS	400×10^6
19.7×10^9 YEARS			FERNS, CONIFERS	300×10^6
19.8×10^9 YEARS	MESOZOIC ERA		FIRST MAMMALS	200×10^6
19.85×10^9 YEARS			FIRST BIRDS	150×10^6
19.94×10^9 YEARS	CENOZOIC ERA		FIRST PRIMATES	60×10^6
19.95×10^9 YEARS			MAMMALS INCREASE	50×10^6
20×10^9 YEARS			HOMO SAPIENS	1×10^5

Figure 15 MAJOR EVENTS IN THE UNIVERSE'S HISTORY. Ancient events are dated in terms of the red shift because the precise age of the universe is not known. At the ultrahigh velocities characteristic of times close to the big bang the red shift is equal to $(1 + v/c)/(1 - v^2/c^2)^{1/2} - 1$, where v is velocity of the radiation source and c is velocity of light (3×10^8 meters per second).

POSTSCRIPT

Our article described the problem of explaining the regularity of the universe and the role of the particle horizon in restricting the smoothing of irregularities by dissipative processes. Since then the idea of a cosmological "inflation" has been suggested and is discussed in Chapter 11, "The Inflationary Universe," by Alan H. Guth and Paul J. Steinhardt. The chaotic cosmology program, which sought to explain the present regularity of the universe independent of initial conditions, has found its most powerful manifestation in this concept.

The next chapter, "The Large-Scale Structure of the Universe," sheds more light on some of these questions.

The Large-Scale Structure of the Universe

Across billions of light-years space is a honeycomb of galactic superclusters and huge voids. The structure may result from perturbations in the density of matter early in the big bang.

· · ·

Joseph Silk, Alexander S. Szalay and Yakov B. Zel'dovich

stronomers have long recognized that the distribution of matter on a cosmic scale must somehow bear the imprint of a very early stage in the history of the universe. A consistent account of that distribution and its evolution must be developed within the context of the big-bang theory, since there is almost universal consensus among cosmologists and astrophysicists that the big bang provides an empirical framework within which all cosmological issues can be examined. According to the big-bang theory, the universe began as a singular point of infinite density some 10 to 20 billion years ago and pulsed into being in a vast explosion that continues to this day. In the simplest version of the theory the universe expands everywhere uniformly from the singular point. The uniformity of that expansion accounts remarkably well for much important observational evidence: Extragalactic matter recedes from our galaxy at a rate that varies smoothly with its distance, and a cold bath of radiation in the microwave region of the electromagnetic spectrum pervades the sky at a temperature that varies over a few angular degrees to less than one part in 30,000. In spite of these successes, there is compelling evidence that the expansion is not precisely uniform. If it were, matter

would fail to coalesce and the universe would become an increasingly rarefied gas of elementary particles. The stars and the galaxies would not exist.

In order to account for structure in the present state of the universe the big-bang cosmologist must therefore acknowledge some degree of clumpiness early on. Such early inhomogeneities might be smooth and nearly indistinguishable against the homogeneous background; small fluctuations in the curvature of the early universe would take the form of slight compressions or rarefactions of matter and energy from region to region throughout space. The amplitude of the fluctuations must be large enough (that is, the variation from the average density must be great enough) to grow into the currently observed aggregations of matter in the time since the universe began; precisely what that amplitude must be, however, is a matter of much theoretical delicacy. If the initial fluctuations were too large, they would cause variations in the temperature of the microwave background radiation that are not observed. Furthermore, the fluctuations must give rise to the structures of relatively special scale that make up the universe and not to structures of arbitrary size. Stars, galaxies, clusters of galaxies and even superclusters, or clusters of clusters, can now be identified, but at scales of mass larger than that of the supercluster the universe is fairly uniform.

The convergence of cosmology and the physics of elementary particles has recently made it possible to satisfy all these requirements without making any strong assumptions about the early state of the universe. In particular, no appeal is made to any special scales or patterns of mass and energy at the outset of the expansion, and no new forces are invoked. What is assumed is that soon after the beginning of the big bang there were small variations in the density of matter and energy everywhere in the universe. The variations were the result of superposing low-amplitude, sinusoidally varying fluctuations in the density at every possible wavelength, or scale of length; the amplitudes of the fluctuations were distributed randomly, and so the resulting variations in density were random and chaotic. Thereafter the present structure of the universe could have evolved according to reasonably well-understood principles of physics.

As the universe expanded, the random, free streaming of elementary particles in all directions suppressed all the initial fluctuations below a critical size: the only fluctuations that survived were those that compressed or rarefied masses at least 10^{15} to 10^{16} times the mass of our sun. Gravity then caused some of the compressed masses to contract preferentially along one of the three spatial axes. The initial spectrum of fluctuations thereby gave rise to gigantic, irregular clouds of gas that resembled flattened pancakes. Where the pancakes intersected, long, thin filaments of matter took shape. Some of the clouds remained intact; others broke up to form galaxies and clusters of galaxies. The emergence of an appropriate characteristic scale for the fluctuations was first explained by one of us (Silk). The gravitational formation of the thin layers of matter was discovered by another of us (Zel'dovich). We shall refer to this model as the pancake theory.

The pancake theory in its present form is a tale of two objects at the extremes of physical scale. One is an astronomical system large enough to fill 10^{23} cubic light-years of space; the other is the neutrino, a weakly interacting elementary particle that is almost vanishingly small. If the pancake theory is to be confirmed, both objects must be observed and a nonzero mass must be assigned to the neutrino. Since the two masses, if they both exist, span 80 orders of magnitude, extraordinary procedures are needed to measure them from our own vantage of middle dimensions.

Remarkably, the existence of the required scale of the astronomical system has recently been verified, and its mass offers tantalizing evidence that the pancake theory is on the right track. Systematic measurements of distance for several thousand galaxies have been carried out by determining the red shift in their spectra: the displacement of spectral lines toward the long-wavelength end of the electromagnetic spectrum. The red shift is a Doppler effect, caused by the recession of a distant galaxy from our galaxy. The velocity of the recession can be calculated from the red shift according to a simple mathematical formula, and the distance of a galaxy varies directly with its recessional velocity because the universe is expanding. A red-shift measurement combined with the coordinates of a galaxy on the surface of the sky enables the astronomer to fix the galaxy in space. Three-dimensional maps of the distribution of galaxies have thereby been worked out. Figure 16 illustrates just such a map. The distribution of 400,000 galaxies across 100 degrees of the sky is mapped by a computer program designed to enhance the filamentary structure that is perceptible in other maps. The number of galaxies within each pixel, or small square area of the map, is indicated by a color code: black pixels represent areas having the least number of galaxies, and an increasing density of galaxies is indicted by increasingly lighter shades of brown through white. Green and red pixels correspond to local ridges and peaks in the distribution. The green pixels designate regions where the galaxy count is greater than it is in all the adjacent pixels in two or three directions, either horizontal, vertical or diagonal. Red pixels designate regions where the count reaches a local maximum in all four directions. The green filamentary ridges may correspond to superclusters of galaxies in space; if they do, they extend across 100 million light-years.

The maps show features quite unlike those of most other astronomical objects: the galaxies are concentrated in enormous sheets and filamentary structures whose greatest dimension, roughly 100 million light-years, is an order of magnitude larger than its lesser dimensions. Such a structure can include as many as a million galaxies; its mass is on the order of 10^{16} suns. Moreover, within each structure the galaxies are not evenly distributed: one can distinguish more densely populated clumps and strings, many of them at the intersection of two

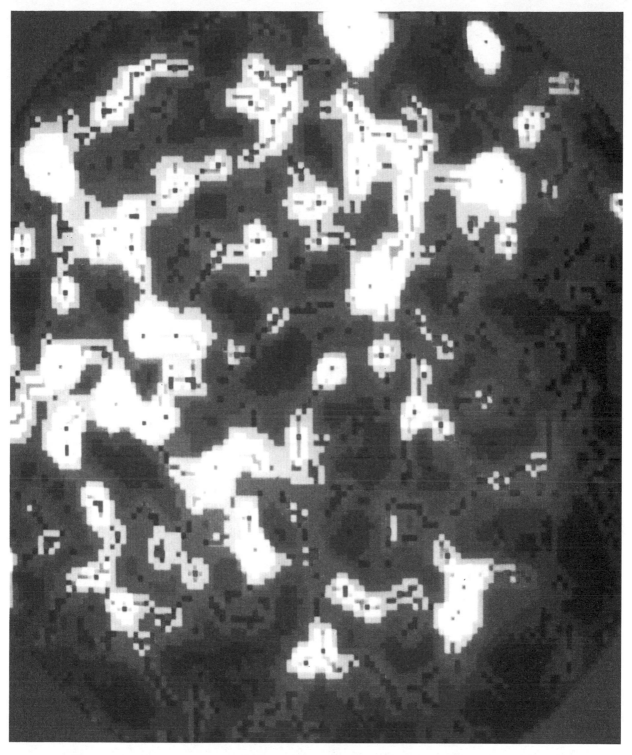

Figure 16 DISTRIBUTION OF 400,000 GALAXIES across 100 degrees of sky. The number of galaxies in each pixel is indicated by a color code. The map is centered on the northpole of the galaxy in the constellation Coma Berenices. (Map by J. E. Moody, E. L. Turner and J. R. Gott III based on a survey by C. D. Shane and C. A. Wirtanen.)

sheets. Finally, interspersed among the largest structures are huge voids, virtually free of galaxies, that are between 100 and 400 million light-years across [see "Superclusters and Voids in the Distribution of Galaxies," by Stephen A. Gregory and Laird A. Thompson; SCIENTIFIC AMERICAN, March, 1982; Offprint 3120]. Much of this picture is based on the work of Jaan Einasto of the Tartu Observatory in the Estonian S.S.R.

The detection of a massive neutrino is much more problematic. Several years ago theoretical physicists assigned to the neutrino a rest mass of zero, but some more recent theories of elementary particles suggest the neutrino does have a small mass. Several kinds of experiment are under way seeking to detect it. In the most direct method the mass can be inferred if certain variations are found in the decay rate of radioactive isotopes. In 1980 Valentin Lubimov, Evgeny Tretyakov and their colleagues at the Institute of Experimental and Theoretical Physics in Moscow measured the decay rate of tritium, the radioactive isotope of hydrogen. At that time they reported results consistent with a small but nonzero neutrino mass, between 14 and 46 electron volts, which is less than a ten-thousandth the mass of the electron. Recently the same investigators have confirmed their findings and narrowed the limits of error: they now report a neutrino mass of from 20 to 40 electron volts.

Unfortunately there has been no independent verification of these results, and so there remains no general consensus on the question of neutrino mass. A second kind of experiment, pioneered by Ettore Fiorini of the University of Milan, is based on the rate of a mode of radioactive decay called double-beta decay that is observed in certain isotopes. Fiorini has reported the neutrino mass can be no greater than 10 to 20 electron volts, based on the decay rate of the isotope germanium 76. The method is less direct than the measurement of tritium decay; the results of Fiorini's experiment can be interpreted as a measure of neutrino mass only if it is assumed that the neutrino is its own antiparticle. On the other hand, if the neutrino and the antineutrino are distinct, the double-beta decay of germanium 76 is modified and a value for the neutrino mass cannot be inferred.

A third method of detecting neutrino mass was first proposed by Bruno M. Pontecorvo of the Joint Institute of Nuclear Research at Dubna in the U.S.S.R. The method exploits the fact that there are three kinds of neutrino: the electron neutrino, the muon neutrino and the tau neutrino. If the three kinds of neutrino have mass, if the three kinds can appear with varying probability and if the difference between the squares of the masses of any two kinds of neutrino is not equal to zero, quantum mechanics implies that the three kinds of neutrino could oscillate, or freely exchange their identities. Since the oscillations would cause the population of one kind of neutrino to vary with time, the oscillations should be detectable as a change in the population of, say, electron neutrinos along the path of a neutrino beam. Several such experiments have been done in the past few years, first in 1980 by Frederick Reines and his colleagues at the University of California at Irvine and later by Felix H. Boehm and his colleagues at the California Institute of Technology and by other workers. At this writing no experimental group has reported unambiguous evidence for neutrino oscillations. Unfortunately the absence of oscillations could merely indicate that the difference between the squares of the masses of two kinds of neutrino is zero; a failure to detect oscillations is therefore consistent with a finite, or nonzero, neutrino mass.

The prevailing attitude among physicists is that the experimental results do not yet warrant a firm conclusion about the mass of the neutrino. Nevertheless, if the evidence for mass is accepted, the cosmological consequences are far-reaching. More than a decade ago, following an early suggestion by Semyon Gershtein of the Serpukhov Institute of Physics in the U.S.S.R. and one of us (Zel'dovich), György Marx and one of us (Szalay) at Eötvös University suggested that massive neutrinos could make a dominant contribution to the mass and evolution of the universe as a whole. This suggestion was made concurrently by Ramanath Cowsik and John McClelland of the University of California at Berkeley. A massive neutrino would also lead inevi-

Figure 17 SMALL PERTUBATIONS IN THE DENSITY of matter and energy shortly after the big bang can be understood as wavelike fluctuations of the density around an average value, randomly distributed over all wavelengths. Such a fluctuation is shown in cross section by the computer-generated image at top; yellow, green and blue mark relatively compressed regions, and orange, red and purple are increasingly rarefied regions. In the image at bottom all fluctuations encompassing a mass of less than 10^{16} times that of our sun have been filtered out by the interaction of matter and radiation in the early stages of the universe. (Computer simulation by S. Djorgovsky.)

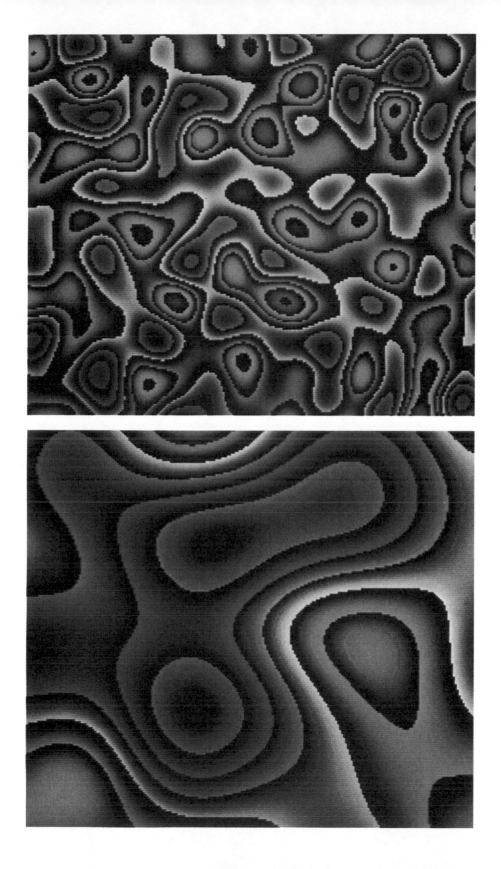

tably to pancakelike structures on large scales. Before we discuss this effect, however, it will be useful to describe an earlier version of the pancake theory, a theory that ultimately failed certain critical observational tests but that has given rise to the current more successful theory.

Astrophysicists believe they have a fairly sound understanding of the physical processes that must have taken place after the first few milliseconds of the big bang. The energies of particles colliding with one another at that time were no greater than the energies typically achieved in small particle accelerators, and so a picture of the early universe emerges when one considers a dense fluid of particles whose individual properties are well known.

By far the most abundant particles in the fluid were the photon, the electron and the three species of neutrino; only relatively small numbers of protons and neutrons were left over from annihilations that took place in an earlier epoch. The electrons and the neutrinos remained in close contact throughout most of the first second and were continuously created and annihilated. Frequent collisions among them guaranteed that energy was distributed randomly throughout the fluid; in other words, the particles were maintained in thermal equilibrium. As the universe expanded, the density of the particles decreased and collisions became less frequent. Because the energy of a photon varies inversely with its wavelength, the average energy of the photons decreased as their wavelengths expanded with the rest of the universe, and so the universe began to cool.

Recent theoretical investigations that have sought to unify the fundamental forces of nature can now peer even farther back than the first millisecond into the history of the universe. The theories are called grand unified theories because they attempt to understand the electromagnetic force, the weak nuclear force and the strong nuclear force as distinct low-energy manifestations of a single, underlying phenomenon. (Gravity, the fourth fundamental force, has not yet been incorporated.) The energy density at which the three forces become indistinct corresponds to the energy density of the universe only 10^{-35} second after the start of the big bang. The early universe has therefore come to be regarded as a laboratory for testing the predictions of grand unified theories.

One prediction of the theories is that if the density of matter must fluctuate in the early stages of expansion, the density of the photons, or radiation, must fluctuate also. Nevertheless, the ratio of the density of matter to the density of radiation must always remain the same. According to the general theory of relativity, energy and matter are equivalent as a source of gravity and determine the geometry of space-time. A fluctuation in the density of mass and energy therefore causes a fluctuation in the gravitational field, which is equivalent to a fluctuation in the curvature of space-time. The comprehensive theory of such fluctuations in the expanding universe, treated within the framework of the general theory of relativity, was developed in 1946 by Eugene M. Lifshitz of the Institute for Physical Problems in Moscow.

It seems reasonable to assume that fluctuations must have existed in the early universe over a wide range of possible scales. We make the assumption primarily for reasons of parsimony: it would seem arbitrary and entirely fortuitous if the initial fluctuations were to single out regions only of, say, galactic scale. There is an upper limit, however, to the size of fluctuation that can be perceived by any observer at a given time. That limit is the spatial horizon of the observer, which is a sphere, centered on the observer, whose radius is equal to the distance light can travel in the time since the start of the big bang. Figure 18 shows the horizon of an observer as a circle at the base of a cone. The expansion of the universe since the beginning of the big bang is indicated by the divergence of two galaxies with time. In the early stages of expansion the galaxies recede from each other at an apparent velocity that exceeds the speed of light, and so at this epoch there has not been enough time for an observer in one of them to see the other galaxy at a previous stage (a). Because the expansion is slowing, however, light from an early stage of one of the galaxies will eventually reach an observer in the other galaxy (b). The edges of the cone are the paths of light signals in space; since no signal can propagate faster than light, they represent a spatial limit to causal interaction at any given time as well as a limit to observation. Fluctuations undetectable in the early universe become detectable later on, because they begin to be encompassed by the observer's ever widening horizon.

Once a fluctuation is within an observer's horizon it is adequately described by classical, or nonrelativistic, gravitational theory. There it takes the form of an observable perturbation in the density of the fluid. There are two competing effects on

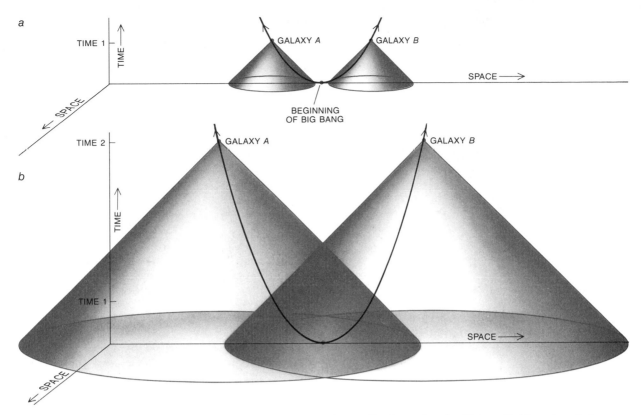

Figure 18 HORIZON OF AN OBSERVER grows with time and encompasses an increasingly great fraction of the universe. The horizon is a sphere, centered on the observer, whose radius is equal to the distance light can travel since the beginning of the big bang. In the figure only a circular slice of the spherical horizon is shown in the space plane.

any parcel of matter and radiation moving together: gravity tends to collapse the parcel, and pressure caused by the chaotic motion of the particles and radiation tends to disperse the parcel in space. On large scales gravity always wins. Pressure cannot resist the collapse, so that particles are attracted to the regions of highest density. Moreover, once gravitational collapse begins, the accreted mass can attract more distant mass and radiation, and so any initial instability is amplified. Matter accumulates in some regions and becomes rarefied in others.

If the soup of particles and radiation that make up the early universe is considered to be an ideal gas, the effects of a density fluctuation superposed on the gas are straightforward. Any local compression in the density over a sufficiently large mass will trigger gravitational instability and lead to incipient collapse. On smaller scales, however, gravity is not strong enough to overcome the increase in the pressure of the gas caused by the increase in density. The compressed parcel of gas will therefore rebound and become rarefied, and the fluctuation will

propagate exactly like a sound wave, that is, by the periodic compression and rarefaction of the medium through which it travels.

Most sound waves in the air die out in a few tens of meters because the particles that make up the pressure waves are scattered and their coherent motion is dissipated as heat. In a similar way sound waves in the cosmic medium set up by the fluctuations lose their energy and die away at all but the longest wavelengths. Furthermore, the particles and photons in the early universe are much too densely packed to be treated as an ideal gas. In the first 300,000 years of the big bang the photon radiation was energetic enough to keep all matter ionized. Photons outnumbered electrons by a factor of about 100 million; the free electrons, which would later be bound to atomic nuclei, were therefore under constant bombardment by the photons, freely scattering the photons and being scattered by them. The result was a thick, viscous fluid of electrons and photons in which no particle could travel very far without being scattered.

The scattering of the free electrons by the radiation makes any displacement of the electrons through the radiation much like moving through viscous, cold molasses. The viscosity of the fluid thereby inhibits the growth of gravitational instabilities that might be caused by the accretion of matter alone. Moreover, as in the ideal gas, the large pressure of the radiation keeps matter and radiation from collapsing together under the pull of gravity on all but sufficiently large scales. The remaining fluctuations within the viscous fluid, that is, the ones that survive gravitational collapse, can be regarded as sound waves.

As we have mentioned, grand unified theories require that the ratio of the density of photons to the density of matter remain always the same: in the compression phase of a fluctuation the compression of the photons must therefore match the compression of the particles with mass. If the distance covered by a photon in the time since the start of the big bang is greater than the distance across a compressed region of a fluctuation, however, the photon will in effect not take part in the compression but will instead dissipate its share of the energy of the fluctuation. Since the photons greatly outnumber the particles with mass, they carry almost all the energy of the fluctuations, and so fluctuations on a scale smaller than the average radial displacement of a photon in the time since the start of the big bang are damped out.

The path of the photon can be compared to the path of a drunkard staggering away from a lamppost along any direction with equal probability. In order to wander a distance away from the lamppost that corresponds to N steps when sober, the drunkard must take N^2 steps. Similarly, the photon must be scattered N^2 times in order to travel a radial distance equal to the distance it would travel if it were streaming freely. In spite of their being scattered by the electrons, the photons diffuse radially outward through the medium so fast that they dissipate the energy of all but the largest fluctuations. By the time the universe cools enough for atomic nuclei to capture the free electrons, the photons have diffused through a region of the universe whose mass is about 10^{14} times that of our sun. All initial fluctu-

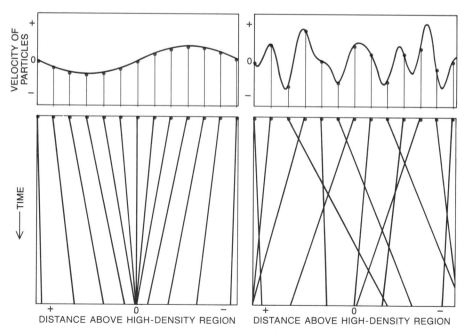

Figure 19 DISTRIBUTION OF PARTICLE VELOCITIES from a fluctuation with a single wavelength. Particles are gravitationally attracted to the regions of highest density and their velocities depend on the distance from such a region (*upper left*). The velocity of each particle (*lower left*) is given by the slope of the trajectory line. The trajectories tend to converge and form regions of enhanced density. If the fluctuations are random over all the wavelengths (*upper right*), the trajectories do not converge (*lower right*).

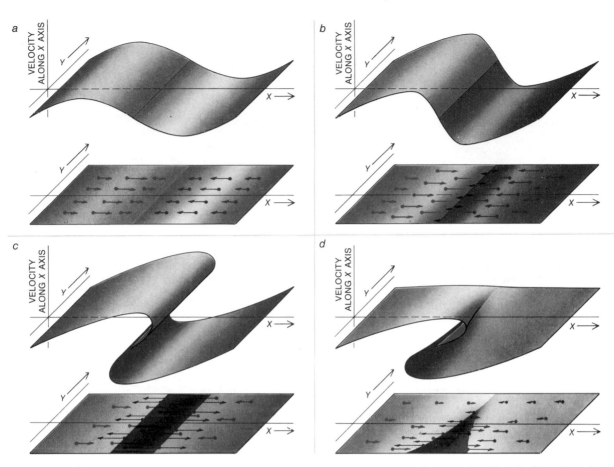

Figure 20 DISCONTINUITIES IN THE DENSITY OF MATTER can result from continuous deformation of matter during gravitational collapse. Particles in the lower plane in each succeeding panel, with velocities given by the upper plane distributions, result in the density cusps.

ations that encompass a mass smaller than this amount are therefore erased.

When the electrons finally combine to form atoms, the matter and the radiation cease to interact and the radiation streams off independently of the atoms. The viscosity of the fluid drops abruptly, and the fluctuations that have survived the previous era of radiation-dominated interactions are no longer inhibited from being amplified. Thereafter gravitational instability proceeds with full vigor.

The sudden absence of radiation pressure has a dominant effect in determining the shape and structure of the first objects to form. Thermal pressure always acts isotropically, or equally in all directions, and so if the radiation pressure had remained com-

parable in strength to gravity, all collapsed objects would have had nearly perfect spherical symmetry. Anisotropies develop because the pressure is completely negligible up to the last moments of collapse.

Figure 20 illustrates this process. Discontinuities in the density of matter can result from continuous deformations of the medium during gravitational collapse. In each frame of this schematic sequence the lower plane represents two directions in space; for simplicity only the motions of particles in the plane and parallel to the x axis are considered. The particles move toward a central axis of the plane; the length of each arrow indicates the velocity of a particle at the position corresponding to the tail of the arrow (*a, b*). A sheet of particles moving to the right (*colored arrows*) crosses the central axis without colliding with the particles moving to the left (*black*

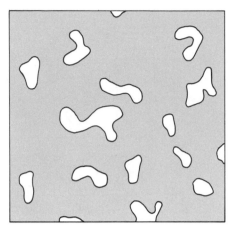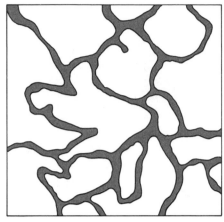

Figure 21 STRUCTURE OF THE FLUCTUATIONS that survive filtering in the early stages of the universe is preserved by the structure of galaxy superclusters and voids. In a nearly uniform universe (*left*), where the probability of rarefaction caused by fluctuation along an axis is one-half, the probability that any region is rarefied along all three spatial axes is one-eighth. Such regions (*shown in white*) occupy one-eighth of the volume of the universe. Gravity compresses the colored regions (seven-eighths of the mass) to form a cobweb of filaments that eventually fill only one-eighth of the volume of space (*right*).

arrows). The motions set up two discontinuities in the density of the medium, one on each side of the central axis (*c*). In almost any real collapse the motions of the two sheets of particles are not exactly symmetrical and the two discontinuities end in a cusp (*d*). The same density distributions can be understood as a special case of a more general phenomenon described by the branch of mathematics called catastrophe theory. If the motions of the particles are plotted in a phase space, that is, in a three-dimensional space where the vertical axis represents the velocity in a direction parallel to the *x* axis, the interaction of the two sheets of particles is represented by a twisting or undulating surface (*upper plane in each frame*). The density of the particles at any point is then given by the "shadow" cast by the deformed surface on the original *x*-*y* plane.

Because of the lack of pressure to counteract infall, gravitational instability is highly efficient at sweeping almost all the matter into compressed, high-density regions of space. Consider the following argument. Along any one of the three spatial axes matter can be either compressed or rarefied; assume, for the sake of simplicity, that the probability of the matter being compressed along any one axis is one-half. The fraction of the gas that will not be compressed along any axis is the cube of one-half, or one-eighth. This result has immediate implications for the predicted spatial structure following collapse. At an early stage, when the density is

still nearly uniform, the regions destined to be compressed include about seven-eighths of all matter. These regions surround smaller bubbles of matter that never collapse; the bubbles are destined to become voids. After the collapse the compressed regions occupy only one-eighth of the volume of space; the small bubbles, which carry one-eighth of the matter, expand to fill the remaining seven-eighths of the volume. The topology of the initial state is preserved. The final outcome is a cellular structure formed by thin walls and filaments of compressed matter that enclose huge voids.

The shape of the compressed regions can be predicted from similar considerations. It is most unlikely that any cubical volume of matter destined for collapse will form a sphere. Such a collapse would require a match of both direction and magnitude of the fluctuations along all three components into which any arbitrary collapse can be resolved. It is much likelier that the cube would collapse first along one randomly selected axis, and it would collapse or expand more slowly along the other two axes. The ensuing distribution of matter is highly anisotropic. Since the mass inside the initial cubical volume does not change as both the thickness and the volume of the cube decrease, the density becomes extremely high and a flat pancake is formed.

At first the pancakes develop in isolated regions, but they soon grow into thin sheets that intersect and form the cellular structure. Numerical simulations of the collapse done with the aid of large computers suggest the universe has only recently acquired a cellular structure. In the future, as larger clumps of matter form, the cellular structure is expected to disappear. Hence it is only during an intermediate stage of cosmic evolution that the initial curvature fluctuations are reflected by the structure of matter. The observational evidence shows that from the perspective of the evolution of large-scale structure the universe is neither very young nor very old.

There are two major difficulties with the pancake theory as we have described it so far. First, remember that the smallest fluctuations to survive the radiation era encompass a mass of 10^{14} suns. Structure in the distribution of galaxies, however, exists at much larger scales. Numerical simulations favor a theory in which the smallest fluctuations emerging from the radiation era are on a scale of 10^{15} or 10^{16} suns.

The second difficulty is more serious. Because the microwave background radiation has propagated freely ever since the photons and the electrons ceased to interact, the variation in the temperature of the radiation across the sky reflects primordial inhomogeneities in the distribution of matter. At the time the original pancake theory was formulated the upper bound for the temperature variation over the entire sky was about one part in 1,000. Accordingly it was thought that matter inhomogeneities in the early universe could have been as great as a third of the temperature variation, or one part in 3,000. Recently more stringent limits on the variation of the radiation temperature have been set by Francesco Melchiorri and his co-workers at the University of Florence and the University of Rome, and by Yuri N. Parijskijin of the Pulkovo Observatory in Leningrad. The new upper bound is a variation of less than one part in 30,000 over an angle of six degrees.

The fluctuations required by the original version of the pancake theory were consistent with the earlier estimate of temperature variations, but the agreement with the new estimate is only marginal. Moreover, if the overall density of matter and energy in the universe is so small that the present expansion will continue forever, the agreement is lost. On a cosmic scale the force of gravity would have been so weak in recent epochs that fluctuations must have completed their growth and collapsed at a much earlier time, when the density of matter was much greater than it is today. The amplitude of such fluctuations, however, would have been much too large to be compatible with the uniformity in the microwave background. On the other hand, if the initial fluctuations had been small enough to be compatible with the radiation background, the birth of galaxies would have become practically impossible.

If the universe is dense enough for the amplitude of the fluctuations to be marginally reconciled with the uniformity of the background, another problem arises. The density cannot then be accounted for solely by the total mass of bright matter, visible as stars, nebulas, galaxies and the like. Instead the universe must be made up predominantly of dark matter. This inference is not a new one. Studies of the rotation of our galaxy and that of other spiral galaxies have shown that the rotational velocities of stars on the periphery of a galaxy are not consistent with Kepler's laws. These laws state that the rotational velocity should decrease with increasing distance from the center of a galaxy, just as the orbital velocity of a planet decreases with its distance from the sun. Peripheral stars, however, do not slow down; their rotational velocities are roughly constant and independent of their distance from the galactic center. P. James E. Peebles and Jeremiah P. Ostriker of Princeton University and Einasto simultaneously suggested the dilemma would be resolved if halos of invisible matter make up the bulk of the mass of spiral galaxies. An indirect argument suggests dark material may be present in even greater quantities within groups and clusters of galaxies. Such systems would fly apart in an unaccountably short time if it were not for the gravitational attraction of dark matter. It is estimated that dark matter may comprise 90 percent of the mass of the universe.

A new component of the universe was badly needed to salvage the pancake theory, and a source of dark matter was needed to account for the motions of galaxies. A natural candidate for both purposes was the neutrino, although certain other exotic but still undetected particles, such as a massive photino or a massive gravitino, might serve the same cosmological function. Theories of elementary particles predict that in the first millisecond of the big bang a wide variety of weakly interacting particles were maintained in thermal equilibrium. Many such particles could still survive, and provided they

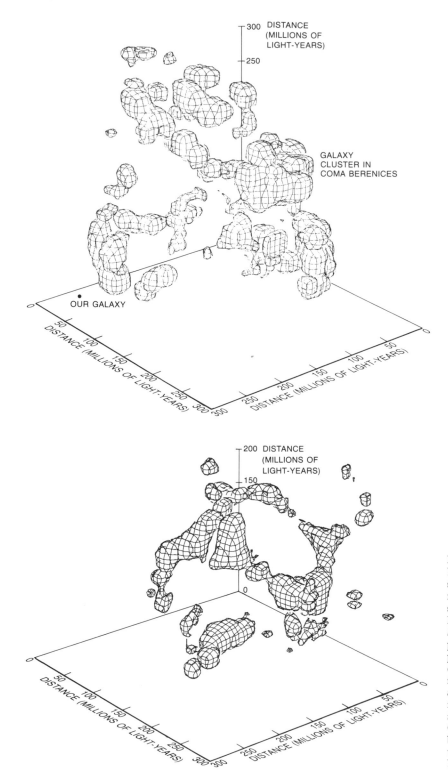

Figure 22 CONTOURS OF HIGH DENSITY plotted in three dimensions for all galaxies brighter than magnitude 14.5 in the northern sky within about 250 million light-years of our galaxy (*upper illustration*) are similar to computer-simulated predictions of the pancake model, assuming the clustering of matter arises from fluctuations no smaller than those able to survive the smoothing effect of a massive neutrino (*lower illustration*). (Real distribution by C. Frenk and S. White from a map by M. Davis and colleagues. Simulation by Frenk, White and Davis.)

are stable they could have far-reaching implications for cosmology. Since the neutrino mass can be measured experimentally, in the remainder of this discussion we shall refer to the neutrino. Nevertheless, even if the neutrino turns out to have no mass, the pancake theory is not disproved.

Remember that in the first second of the big bang the primordial soup included an abundance of neutrinos. Even today the ratio of photons to all three varieties of neutrino is only 11 to 9. Neutrinos, unlike protons, electrons and even photons, interact so weakly with other particles that they begin to stream freely through the fluid long before the photons do. Hence the neutrinos, which initially move at the speed of light, can travel farther than the photons in the early stages of the universe. By the end of the radiation era the neutrinos have dissipated fluctuations on a larger scale than photons alone could have done.

A massive neutrino cannot continue indefinitely moving at the speed of light. When the energy density of the photons falls below the energy that corresponds approximately to the rest mass of the neutrino, the neutrino begins to slow down and move at a speed appropriate to its energy. If the mass of the neutrino is 30 electron volts, the slowdown will begin well before the capture of the free electrons by atomic nuclei. The capture must wait until the background energy is reduced to .1 electron volt, the energy at which hydrogen is ionized by the dense fluid of photons. Although the neutrinos continue to erase the fluctuations as they slow down, they become increasingly susceptible to being trapped by large fluctuations that have not yet been smoothed out. Richard Bond of Stanford University and one of us (Szalay) have estimated the maximum scale over which the neutrinos can freely stream before they are trapped, and consequently the minimum scale over which the fluctuations are not erased. The scale corresponds to a present distance of 100 million light-years and a mass of 10^{15} to 10^{16} suns. The agreement with the size and mass of the galaxy superclusters that are now observed is striking.

How can such fluctuations be compatible with the observed uniformity of the background radiation? The neutrinos cease their erasure of the curvature fluctuations before the end of the radiation era, but unlike the electrons their motions are not inhibited by the viscosity of the fluid. Neutrinos collide so rarely with photons or electrons that they are not subjected to viscous drag. Gravitational instabilities among the neutrinos can accordingly begin to develop before the end of the radiation era, and so they can grow over a much longer time than the fluctuations of ordinary matter can. The initial amplitude of the neutrino fluctuations needed to account for the present inhomogeneities of matter could therefore have been much smaller than the amplitude of the fluctuations needed in a mixture of radiation and ordinary matter. With massive neutrinos the variation in the temperature of the background radiation required to generate the observed aggregations of matter is reduced by an order of magnitude or more. Thus theory and observations can be reconciled.

The new version of the pancake theory leads to a natural explanation for the origin of the dark matter in the universe. The initial collapse of a pancake distributes most of the neutrinos widely because most of them are accelerated by the collapse to large velocities, on the order of 1,000 kilometers per second. Such neutrinos are destined to fill the dark regions of intergalactic space. Other neutrinos, however, move more slowly because they are initially closer to the central plane of the pancake and do not undergo large accelerations. The thin layer of gas in the vicinity of the central plane condenses and breaks up to form protogalaxies. The slow-moving neutrinos are gathered together by aggregates of ordinary matter, and the matter near the center of the protogalaxy continues to condense and form stars. Neutrinos at the periphery of the protogalaxy, however, are gravitationally shared and never condense; they become the dark matter in the galactic halo.

A more detailed theory of galaxy formation within the context of the new pancake theory is now under development. As a pancake collapses, the neutrino component of the collapsing gas passes through the central plane of the pancake without interaction. The density distribution of the neutrinos acquires sharp discontinuities, some of which can be identified with rich clusters of galaxies. Vladimir I. Arnold, a mathematician at Moscow State University, has recently collaborated with astrophysicists on the problem, and he has identified such discontinuities in the overall density distribution with certain elementary structures in the branch of mathematics called catastrophe theory.

THREE BILLION YEARS

FIVE BILLION YEARS

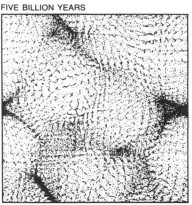

NINE BILLION YEARS

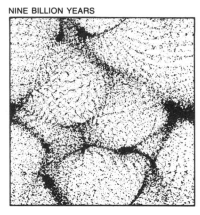

Figure 23 COMPUTER SIMULATION of the evolution of large-scale fluctuations in the density of matter and energy with time, assuming the existence of neutrinos with a non-zero mass. Pancakes and filaments develop as a result of gravitational collapse, and matter in the remaining regions of space become increasingly rarefied. The resulting structure resembles the distribution currently observed of superclusters and voids. (Simulation by G. Efstathiou.)

The pancake theory, as it has now been modified, offers deep insight into the character and origin of the contemporary structure of the universe. It is based on well-known physical principles and on plausible assumptions about the conditions in the very early universe. As a theory of the origin of large-scale structure it is by no means unique, although it now appears that both theory and observation point in the general direction we have outlined. Nevertheless, there are many important questions that must be resolved before the theory can be considered firmly established.

Given the confirmation of the theory, there are two main lines along which it must be developed. First, one must address the finer structure of the universe, the formation of the first generation of stars from a primordial gas that was totally devoid of the heavy elements. Second, one must ask how the initial conditions assumed by the pancake theory arose from even earlier epochs in the history of the universe. Attempts are now in progress to show how the small-amplitude fluctuations required by the pancake theory could result from earlier and much less well-understood phenomena. These attempts are based on theories not yet settled, but preliminary results leave room for optimism that by the end of the century we shall possess a coherent theory of the universe.

POSTSCRIPT

In its simplest form, the neutrino-dominated universe is now recognized to be untenable, for if there are no galactic-size fluctuations prior to the epoch of pancaking, an excessively clumpy universe is created. Galaxies must have formed by the epoch of the most remote quasars, those at a red shift of 4. Thus, pancaking must have occurred if we are to adopt a top-down evolution of large-scale structure. Simulations show that with such an early start of the development on large-scale structure, clustering of galaxies today would be greater than observed.

If initial seeds are postulated, fluctuations allow galaxies to form prior to pancaking of the neutrino dark matter. Suggestions for these seeds are for comparable cold dark matter capable of undergoing small-scale clustering. Possibilities include primordial black holes or, most exotic of all, cosmic strings. These leftover line-like topological defects from a phase transition at grand unification symmetry breaking would accrete matter once the universe became matter-dominated. Cosmic strings could drive galaxy formation, and pancaking of the neutrino-dominated universe would subsequently create the dramatic voids, filaments and sheets that demarcate today's observed large-scale galaxy distribution.

SECTION

III

UNIFICATION IDEAS

. . .

A Unified Theory of Elementary Particles and Forces

At a range of 10^{-29} centimeter the world may be a simple place, with just one kind of elementary particle and one important force. If the proposed unified theory is correct, all matter is unstable.

• • •

Howard Georgi

There can be nothing simpler than an elementary particle: it is an indivisible shard of matter, without internal structure and without detectable shape or size. One might expect commensurate simplicity in the theories that describe such particles and the forces through which they interact; at the least one might expect the structure of the world to be explained with a minimum number of particles and forces. Judged by this criterion of parsimony, a description of nature that has evolved in the past several years can be accounted a reasonable success. Matter is built out of just two classes of elementary particles: the leptons, such as the electron, and the quarks, which are constituents of the proton, the neutron and many related particles. Four basic forces act between the elementary particles. Gravitation and electromagnetism have long been familiar in the macroscopic world; the weak force and the strong force are observed only in subnuclear events. In principle this complement of particles and forces could account for the entire observed hierarchy of material structures, from the nuclei of atoms to stars and galaxies.

An understanding of nature at this level of detail is a remarkable achievement; nevertheless, it is pos-

sible to imagine what a still simpler theory might be like. The existence of two disparate classes of elementary particles is not fully satisfying; ideally one class would suffice. Similarly, the existence of four forces seems a needless complication; one force might explain all the interactions of elementary particles. An ambitious new theory now promises at least a partial unification along these lines. The theory does not embrace gravitation, which is by far the feeblest of the forces and may be fundamentally different from the others. If gravitation is excluded, however, the theory unifies all elementary particles and forces.

The first step in the construction of the unified theory was the demonstration that the weak, the strong and the electromagnetic forces could all be described by theories of the same general kind. The three forces remained distinct, but they could be seen to operate through the same mechanism. In the course of this development a deep connection was discovered between the weak force and electromagnetism, a connection that hinted at a still grander synthesis. The new theory is the leading candidate for accomplishing the synthesis. It incorporates the leptons and the quarks into a single family and provides a means of transforming one kind of particle into the other. At the same time the weak, the strong and the electromagnetic forces are understood as aspects of a single underlying force. With only one class of particles and one force (plus gravitation), the unified theory is a model of frugality.

Leptons and quarks are known to have quite different properties; how can they be consolidated into a single family? The weak, the strong and the electromagnetic forces differ in strength, range and other characteristics; how can they be derived from a single force? The unified theory does not attempt to conceal the differences, but it asserts they are not fundamental. The differences are conspicuous mainly because the universe is now quite cold, so that particles generally have low energy. If experiments could be done at extremely high energy, the unification would become apparent in all its simplicity. Leptons and quarks would be freely interconverted and the three forces would all have the same strength.

The energy needed in order to see the unification of particles and forces in this dramatic form is estimated to be about 10^{15} gigaelectron-volts, or GeV. (One GeV is the energy imparted to an electron when it is accelerated through a potential difference of a billion volts.) This energy exceeds the capabilities of even the largest planned particle accelerators by a factor of 10 trillion, and it is most unlikely such an energy will ever be attained in the laboratory. It may therefore seem that the theory can never be tested, but such is not the case; the theory has definite consequences at readily accessible energies.

First, the theory provides a rationale for several established features of the physical world that have long seemed mysteriously arbitrary. It accounts for the quantization of electric charge: the observation that charge always comes in discrete multiples of a fundamental smallest charge. It gives a value for the relative strengths of the three forces (measured at ordinary laboratory energy) that is in reasonably good agreement with experimental results. It might explain why there is more matter than antimatter in the universe. Equally important, the unified theory predicts new phenomena that cannot be deduced from earlier theories. The most noteworthy of these predictions is the decay of the proton, a particle that had been considered absolutely stable. If the proton can decay, the atom itself is unstable, and all matter is impermanent.

The unified theory is not intended to supplant the established theories of the weak, the strong and the electromagnetic forces. On the contrary, the individual theories are embedded in the structure of the larger one. In order to explain the nature and the origin of the unified theory it is therefore best to begin with the individual component theories, with the forces the theories describe and with the elementary particles on which the forces act.

The apparent differences between the leptons and the quarks are substantial. Six leptons are known (see Figure 24), and among them the electron can be considered prototypical. It has a small mass, equivalent in energy units to about 500,000

	LEPTONS		QUARKS					
THIRD GENERATION	ν_τ	0	t	$+2/3$	t	$+2/3$	t	$+2/3$
	τ^-	-1	b	$-1/3$	b	$-1/3$	b	$-1/3$
SECOND GENERATION	ν_μ	0	c	$+2/3$	c	$+2/3$	c	$+2/3$
	μ^-	-1	s	$-1/3$	s	$-1/3$	s	$-1/3$
FIRST GENERATION	ν_e	0	u	$+2/3$	u	$+2/3$	u	$+2/3$
	e^-	-1	d	$-1/3$	d	$-1/3$	d	$-1/3$

Figure 24 LEPTONS AND QUARKS differ in a number of important properties, including electric charge. Lepton charges are integers, whereas quark charges are fractions. Furthermore, the leptons exist as free particles, whereas the quarks are found only as constituents of composite particles called hadrons. Leptons and quarks are usually divided into three generations; only particles of the first generation have a place in the structure of ordinary matter.

electron volts, and it has one unit of electric charge; by convention the charge of the electron is negative. Two other leptons, namely the muon and the particle designated tau, have the same charge and indeed seem to be identical with the electron in all properties except mass. The muon is more than 200 times as massive as the electron; the tau lepton, which was only discovered several years ago, has a mass almost 3,500 times that of the electron.

The remaining leptons are three kinds of neutrino, which are electrically neutral and have a very small mass (if they have any mass at all). One neutrino is associated with each charged lepton. In addition, for each of the six leptons there is an antilepton with the same mass but with the opposite electric charge. Thus the antielectron (or positron), the antimuon and the antitau particle all have a charge of +1. The antineutrinos, like the neutrinos themselves, have no electric charge.

Whereas the leptons appear as free particles, no one has yet been able to examine a quark in isolation. The quarks are observed only as constituents of the particles called hadrons, a large and diverse class that includes the proton, the neutron, the pi meson and more than 100 other known particles.

There is compelling evidence for the existence of five kinds of quark: they are designated down (d), up (u), strange (s), charm (c) and bottom (b). A sixth kind labeled top (t) has been predicted, but so far it has not been seen. The kinds of quark are generally called flavors; in addition the quarks have another property called color. (Flavor and color are arbitrary labels, which have nothing to do with the sensations of taste or sight.) Each flavor of quark comes in three colors: red, green and blue. The property of color marks a major difference between the leptons and the quarks. The five or six flavors of quark correspond roughly to the six varieties of lepton, but there is no counterpart among the leptons of the quark colors. The distinction has observable consequences. The strong force is an interaction between colors, and since the leptons have no color, they are not susceptible to the strong force.

Another distinctive property of the quarks is their electric charge. The d, s and b quarks have a charge of $-\frac{1}{3}$, whereas the u, c and t quarks have a charge of $+\frac{2}{3}$. The antiquarks, which are labeled $\bar{d}$, $\bar{u}$ and so on, have opposite values of electric charge; hence the $\bar{d}$ antiquark has a charge of $+\frac{1}{3}$ and the $\bar{u}$ antiquark has a charge of $-\frac{2}{3}$. The antiquarks also have opposite colors, namely antired, antigreen and antiblue.

Quarks can be combined in two ways to form a hadron. Either three quarks are bound together, with one quark of each color, or one quark of a given color is joined to an antiquark of the corresponding anticolor. These combinations are said to be colorless, and it turns out they have another significant trait as well. In all the allowed combinations the fractional electric charges of the quarks add to yield an integral total charge; no other combinations (except multiples of the allowed ones) have this property. The proton has the quark composition uud, giving it a total electric charge of $\frac{2}{3} + \frac{2}{3} - \frac{1}{3}$, or +1. The neutron consists of the quarks udd, with charges of $\frac{2}{3} - \frac{1}{3} - \frac{1}{3}$, for a net charge of zero. The positive pi meson is made up of a u quark and a $\bar{d}$ antiquark; the component charges of $+\frac{2}{3}$ and $+\frac{1}{3}$ give a total of +1.

The fact that all atoms are electrically neutral implies that the charge of the proton has exactly the same magnitude as that of the electron, although of course the signs are opposite. For the same reason the charge of the neutron must be exactly zero. It follows from these observations that the charges of the quarks must be exactly commensurable with those of the leptons. For example, the charge of the d quark must be exactly one-third that of the electron and not merely approximately so. This precise relation among particles that appear to be independent is another seemingly arbitrary property one would like to see explained in a unified theory.

It has become customary to classify the leptons and the quarks in three generations. Each generation is made up of a charged lepton, its associated neutrino and two quarks, one quark with charge $-\frac{1}{3}$ and one with charge $+\frac{2}{3}$. The first generation consists of the electron, the electron-type neutrino, the d quark and the u quark. Because the quarks exist in three colors there are eight particles in the generation. All atoms and all ordinary matter can be assembled from these eight particles; the higher generations are observed almost exclusively in laboratory experiments with accelerated particles. In the unified theory the three generations are described independently but in essentially the same way. I shall therefore discuss only the first generation.

Of the three forces that I shall consider here (see also Chapter 5, Figure 38) electromagnetism was the first to be described by an accurate theory; indeed, the accuracy of the theory has never been surpassed. The theory is quantum electrodynamics, or QED, and it was developed over a period of about 25 years culminating in the early 1950's. It has served as a model for theories of the other forces.

The concept of a force is closely connected with that of a charge. Electric charge is the property attributed to a particle that responds to the electromagnetic force, and the amount of the charge determines the response. When two charged particles approach each other, an attraction or a repulsion is set up whose magnitude is directly proportional to the product of the charges. The force is also inversely proportional to the square of the distance between the charges. These two rules constitute Coulomb's law of the electric force. It is important to note that if one of the particles has zero charge, there is no attraction or repulsion; such neutral particles are not directly susceptible to the electromagnetic force.

How strong is the electromagnetic interaction between charged particles? For any given particles the answer depends on the charges and on their distance of separation, but Coulomb's law can provide an absolute answer. Suppose the force between two particles is multiplied by the square of the distance between them; the product is a measure of the strength of the electromagnetic interaction that is independent of the particles' separation, although it does depend on the system of units in which the separation is expressed. Dividing by the speed of light and by Planck's constant (two quantities that are built into the structure of the relativistic quantum-mechanical world) yields a result that is independent of units. The result is a pure or dimensionless number; it has the same value whether measurements are made in grams, centimeters and seconds or in tons, feet and years, provided only that the speed of light and Planck's constant are expressed in the same units as the measurements.

The strength of any given electromagnetic interaction evidently still depends on the size of the participating charges. Thus the interaction would be four times as strong if each charge were doubled. Since electric charge is quantized, however, the interaction of two protons or two electrons has a special role. All particles observed in isolation (that is, all particles except the quarks) have charges that are integral multiples of the proton's charge, so that the proton-proton interaction is a measure of the minimum strength of the electromagnetic interaction. This quantity is called the electromagnetic coupling constant, and it constitutes an absolute measure of the strength of the interaction. Experimental determinations of the coupling constant yield a value of about $1/137$. Because the value is less than 1 the electromagnetic interaction is fairly weak.

It should be emphasized that the quantization of charge is not required by or predicted by quantum electrodynamics; rather it is an experimental fact. The theory would work as well if there were observable particles with fractional charges or even with irrational quantities of charge, such as pi or the square root of 2.

In quantum electrodynamics the interaction of two charged particles, such as two electrons, is related to the exchange of a third particle. The intermediary particle is a photon: a quantum of electromagnetic radiation. The photon is a massless particle that has no electric charge of its own and that moves (by definition) with the speed of light. Describing the electromagnetic force as an exchange of photons avoids the troublesome notion of action at a distance. The interaction is confined to two point-events: the emission and the absorption of the photon. At the same time, however, the description introduces another equally serious problem: the exchange of the photon seems to violate the laws of nature that require energy and momentum to be conserved.

The apparent violation can be illustrated by imagining two electrons held stationary a short distance apart. Since a force could be measured between the electrons, it must be assumed that photons are being exchanged. Ordinarily when a photon is emitted, it carries away part of the energy and momentum of the emitting particle; similarly, when a photon is absorbed, it adds to the energy and momentum of the absorbing particle. In this way the total quantity of energy and momentum in the system is conserved. In the situation being considered here, however, the emitting particle is held stationary, so that its energy and momentum cannot change, and the same is true of the absorbing particle. Evidently the exchanged photon has rather special properties, different from those of the photons that make up sunlight or radio waves. In recognition of the distinction the exchanged photon is called a virtual photon.

The explanation of the peculiar properties of the virtual photon lies in the uncertainty principle introduced into quantum mechanics by Werner Heisenberg. The uncertainty principle does not invalidate the conservation laws of energy and of momentum, but it does allow a violation of the laws to go unnoticed if it is rectified quickly enough. The stationary electrons have the same energy and momentum before the virtual photon is emitted and after it is absorbed; the conservation laws seem to be violated only during the brief passage of the photon. The uncertainty principle states that such an apparent violation can be tolerated if it does not last too long or extend too far.

How long is too long and how far is too far? The answers depend on the magnitude of the apparent violation: the greater the imbalance in energy and momentum caused by the emission of the virtual photon is, the sooner the photon must be reabsorbed. A high-energy virtual photon can survive only briefly, whereas a low-energy one has a long grace period before the books must be balanced. To be explicit, the product of the energy imbalance and the lifetime of the photon cannot exceed Planck's constant. The minimum energy any particle can have is the energy equivalent of the particle's rest mass, and so the maximum range of a virtual particle depends inversely on its mass. The range of the electromagnetic force seems to be infinite, and so the rest mass of the photon is thought to be exactly zero.

The presence of virtual particles greatly complicates the structure of the universe. Because of the virtual particles the vacuum is not mere empty space. A virtual photon can appear spontaneously at any instant and disappear again within the time allotted by the uncertainty principle. Other virtual particles can be created in the same way, including electrically charged particles; the only constraint is that particles with an electric charge must appear and disappear in matched pairs of particle and antiparticle. This process has profound consequences for the theory of electromagnetism.

Consider what happens when a real electron is embedded in a cloud of virtual photons and virtual electron-positron pairs (see Figure 36). The photons have little effect, but the charged virtual particles become polarized: negative virtual charges are repelled by the real negative charge, whereas positive virtual charges are attracted to the real electron. As a result the electron becomes surrounded at close range by a cloud of positive charges, which screen or shield part of the electron's charge.

It follows from this analysis that the "bare" charge of the electron is much greater than the measured charge. Indeed, in quantum electrodynamics the bare charge is assumed to be infinite. The measured charge is merely the finite residue that remains when the shielding charge is subtracted from the bare charge. If the electron's charge could be measured at extremely close range, it would be found to increase as the screening layer was penetrated. A further consequence is that the coupling constant of electromagnetism is not a constant at all but varies with the distance at which charged particles interact with one another. The coupling constant increases (signifying that the electromagnetic interaction becomes stronger) as the range is reduced. The measured coupling constant of about $1/137$ is observed at atomic distances of roughly 10^{-8} centimeter.

Even in the ephemeral realm of the virtual particles there is one conservation law that is never violated: the conservation of electric charge. Because the photon itself is neutral, charge is automatically conserved in the exchange of a virtual photon; no charges are altered. Moreover, when charged matter is created or annihilated, it is always in pairs of particles and antiparticles, so that the sum of the charges after the event remains the same as it was before it.

The conservation of electric charge and the masslessness of the photon are related to a group of symmetries in the mathematical system that describes quantum electrodynamics. The group of symmetries is designated $U(1)$, and so QED is a $U(1)$ theory. $U(1)$ is a term employed in the mathematical theory of groups. The 1 refers to the fact that the

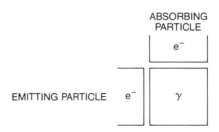

Figure 25 ELECTROMAGNETISM is described by a $U(1)$ symmetry, which signifies that the electromagnetic force cannot change the identity of a particle. The one-by-one matrix is occupied by the photon, which can only transform an electron into another electron.

photon interacts with only one kind of particle at a time. The photon never transforms one kind of particle into another kind. The strong force and the weak force are more complicated in this respect, and the groups that describe them are more complex.

The prevailing theory of the strong force is modeled directly on quantum electrodynamics. The theory is called quantum chromodynamics, or QCD, "chromo-" signifying that the force acts not between electric charges but between color charges. As in QED the magnitude of the force between two charges is proportional to the product of the charges. Particles that have no color charge are not subject to the force. A dimensionless coupling constant defines the intrinsic strength of the interaction. The coupling constant is larger than the constant of electromagnetism, as might be expected of a force that is named strong.

Although QCD is constructed on the same principles as QED, it is a more elaborate theory. The main source of the added complexity is the multiplicity of color charges. Whereas electromagnetism is associated with just one kind of charge, the strong force acts on three colors: red, green and blue. Each color represents a combination of underlying color charges.

There are several ways of analyzing the color charges. The way I shall adopt begins by supposing there are three kinds of color charge. I shall call the charges red minus green $(R - G)$, green minus blue $(G - B)$ and blue minus red $(B - R)$. Each charge can have a value of $+\frac{1}{2}$, $-\frac{1}{2}$ or 0, and each color of quark is distinguished by a particular combination of values. A quark is red if it has an $R - G$ charge of $+\frac{1}{2}$, a $G - B$ charge of 0 and a $B - R$ charge of $-\frac{1}{2}$. A green quark has the color charges $R - G = -\frac{1}{2}$, $G - B = +\frac{1}{2}$ and $B - R = 0$. In a blue quark the three charges are $R - G = 0$, $G - B = -\frac{1}{2}$ and $B - R = +\frac{1}{2}$. The anticolors associated with the antiquarks are formed simply by reversing the signs of all the charges.

Several observations can be made about this distribution of charges. First there are 27 possible combinations of three charges when each charge can have any one of three values. Nevertheless, only quarks with the three combinations that yield the colors red, green and blue seem to exist in nature. Second, this subset of the possible color charges is a highly distinctive one. Each of the observed combinations is arranged so that the sum of the three charges is zero, and the observed combinations are the only ones having this property. (Actually there is one other combination with a total color charge of zero: the combination in which each charge is zero. A particle that has no color charge at all, however, is not a quark.)

The finding that the sum of the three color charges is always equal to zero indicates that one of the three charges is not independent of the other two. If any two of the charges are known, the third can be found by subtraction. Hence it can be concluded that there are really only two varieties of color charge, which are sufficient to specify the three colors completely. It makes no difference which two charges are considered fundamental and which one is discarded; here I shall assume that $R - G$ and $G - B$ charges are fundamental, but I shall often retain the $B - R$ charge for clarity, even though the information it provides is redundant.

A further relation among the charges can be noted. In a state made up of one red, one green and one blue quark the total quantity of each color charge is again zero. In other words, combining the three colors gives rise to a color-neutral state, much as combining an electron and a proton creates a state (the hydrogen atom) that is neutral with respect to electric charge. It is in just this way that color-neutral hadrons such as the proton are formed. A colorless system is also created by combining a color with the corresponding anticolor; since the color charges are then opposite, they cancel exactly. Linking a color with its anticolor is the other formula for making a hadron, on the model of the pi meson. Except for multiples of these combinations (such as a six-quark system that includes two quarks of each color) there is no other way of combining colored quarks so that all the color charges have a sum of zero.

The mechanism by which the strong force is transmitted is comparable to the corresponding mechanism in electromagnetism: the interaction between two charged particles is described as the exchange of a third particle. Again, however, quantum chromodynamics is a more elaborate theory. Whereas QED has a single massless photon, QCD has eight massless particles called gluons. Furthermore, whereas the photon has no electric charge, some of the gluons do carry color charge. The presence of charged carrier particles fundamentally alters the character of the force.

Since the gluons are themselves charged, they not only can transmit the strong force but also can alter the colors of quarks. The emission or absorption of

	RED	GREEN	BLUE
RED	$G_1 + G_2$	$G_{R \to G}$	$G_{R \to B}$
GREEN	$G_{G \to R}$	$G_1 + G_2$	$G_{G \to B}$
BLUE	$G_{B \to R}$	$G_{B \to G}$	$G_1 + G_2$

Figure 26 STRONG FORCE is described by a theory with an *SU*(3) symmetry, in which the couplings of gluons to quarks can be represented by a three-by-three matrix. Any quark color in the left column can be transformed into any of the colors in the top row; the transition is mediated by the gluon specified at the intersection of the column and the row.

photons, in contrast, can never alter the electric charge of a particle. There are nine possible transitions among the quark colors, defined by a three-by-three matrix. For example, a red quark can be transformed into a red quark (the identity transformation), into a green quark or into a blue quark. The three identity transformations (red becomes red, green becomes green and blue becomes blue) make up the diagonal elements of the matrix. It is apparent that the gluons responsible for the identity transitions cannot have color charge or they would alter the quark colors. It might seem there should be three such color-neutral gluons, one neutral gluon for each identity transformation. Since just two independent color charges are needed to specify the three quark colors, however, there are only two color-neutral gluons. I shall designate them G_1 and G_2.

The six remaining transitions between the quark colors do entail changes of color. Each of these transitions is associated with its own gluon, and each of these six gluons does bear a color charge. I shall label the color-charged gluons descriptively. For example, a red-to-green gluon, or $G_{R \to G}$, can be emitted by a red quark, which is thereby transformed into a green quark.

The color charges carried by the gluons can be deduced from the requirement that color charge be conserved. Consider the process in which a quark changes from red to green by the emission of a $G_{R \to G}$ gluon. In the course of the transition the R − G charge of the quark changes from +½ to −½; if the total quantity of charge is to remain unchanged, the gluon must therefore have an R − G charge of +1. In the same way the G − B charge of the quark goes from 0 to +½, so that the gluon is required to carry off a G − B charge of −½. The B − R charge of the quark is transformed from −½ to 0, which again implies a gluon B − R charge of −½. Hence the gluon color charges are respectively +1, −½ and −½. The quark that mediates the reverse transformation, from green to red, must have charges of the same magnitude but opposite sign.

The presence of gluon color charges has a further implication: it automatically ensures that the color charge is quantized. In electromagnetism a photon could in principle be emitted or absorbed by a particle with any quantity of electric charge. Particles with color charge, however, can interact by exchanging gluons only if the charges are separated by intervals that are multiples of ½. It can also be shown that the color charges of the system must be symmetrical about zero, that is, the total of all the positive charges and the total of all the negative charges must be equal in absolute value.

The quantization of the color charge can be demonstrated in another way. Any system of particles with color can be "built" out of the simplest such system: the triplet made up of a red, a green and a blue quark. The triplet of antiquarks can be formed by combining the quarks in pairs. I do not mean to suggest that a physical antiquark consists of two quarks in a bound state. Nevertheless, all the color properties of the antiquark are correctly given by such a synthesis. Note that a red quark has R − G, G − B and B − R color charges of +½, 0 and −½ respectively. An antired antiquark must have the opposite charges: −½, 0 and +½. These are exactly the values found by adding the charges of a green quark (−½, +½ and 0) and a blue quark (0, −½ and +½). Hence antired is in some sense equivalent to the sum of green and blue. Likewise antigreen consists of red plus blue and antiblue consists of red plus green. This remarkable correspondence is a simple consequence of the way the color charges are distributed in the quark triplet. Because the total charge of the triplet is zero, the sum of any two charges must equal the negative of the remaining charge.

The gluons can be constructed in a similar way

COLOR CHARGES

		R−G	G−B	B−R	
QUARKS	RED	+½	0	−½	=0
QUARKS	GREEN	−½	+½	0	=0
QUARKS	BLUE	0	−½	+½	=0
		=0	=0	=0	
GLUONS	G_1	0	0	0	=0
GLUONS	G_2	0	0	0	=0
GLUONS	$G_{R \to G}$	+1	−½	−½	=0
GLUONS	$G_{G \to R}$	−1	+½	+½	=0
GLUONS	$G_{G \to B}$	−½	+1	−½	=0
GLUONS	$G_{B \to G}$	+½	−1	+½	=0
GLUONS	$G_{R \to B}$	+½	+½	−1	=0
GLUONS	$G_{B \to R}$	−½	−½	+1	=0
		=0	=0	=0	

Figure 27 COLOR CHARGES OF QUARKS AND GLUONS. Each of the quark colors red, green and blue is defined by a combination of the three charges, which add up to zero. For a triplet of quarks made up of one quark in each color the sum of the three charges is also zero. Six of the gluons have color charges, with just the values needed to convert a quark from one color into another.

from a quark and an antiquark, although again no physical building process should be inferred. The red-to-green gluon, with charges of +1, −½ and −½, can be imagined as a compound of a red quark (+½, 0 and −½) and an antigreen antiquark (+½, −½ and 0). Of course, the antigreen antiquark can be further decomposed into a red quark and a blue quark, so that the red-to-green gluon has the color properties of two red quarks and a blue quark.

There is yet another consequence of giving the gluons color charges. As explained above, an electron in a vacuum becomes surrounded by a cloud of virtual photons and virtual electron-positron pairs; the charged virtual particles become polarized and shield a part of the electron's bare charge. By the same mechanism a quark in a vacuum becomes enveloped in a cloud of virtual gluons and virtual quark-antiquark pairs, but the result is quite different (see Figure 36). The cloud of virtual quarks and antiquarks is polarized in the usual way, with antiquarks clustered near the real color charge and tending to screen it. The virtual gluons, however, have the opposite effect. The predominant color charge of the gluons near the quark is the same as the charge of the quark itself. Moreover, the virtual gluons are more numerous than the virtual quarks, so that the influence of the gluons is the stronger one. The result is that the charge of the quark appears to be spread out in space, and the effective charge diminishes as the quark is approached more closely.

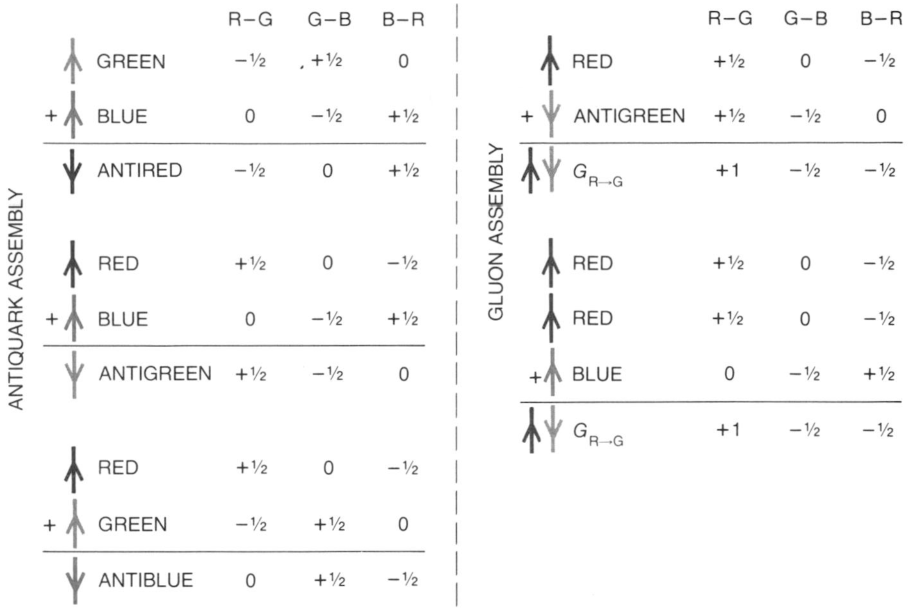

ANTIQUARK ASSEMBLY

		R−G	G−B	B−R
	GREEN	−½	+½	0
+	BLUE	0	−½	+½
	ANTIRED	−½	0	+½
	RED	+½	0	−½
+	BLUE	0	−½	+½
	ANTIGREEN	+½	−½	0
	RED	+½	0	−½
+	GREEN	−½	+½	0
	ANTIBLUE	0	+½	−½

GLUON ASSEMBLY

		R−G	G−B	B−R
	RED	+½	0	−½
+	ANTIGREEN	+½	−½	0
	$G_{R\rightarrow G}$	+1	−½	−½
	RED	+½	0	−½
	RED	+½	0	−½
+	BLUE	0	−½	+½
	$G_{R\rightarrow G}$	+1	−½	−½

Figure 28 ASSEMBLY PROCEDURE for antiquarks and gluons predicts their color properties by supposing they are combinations of quarks of the fundamental color triplet. Any antiquark can be "built" by adding the color charges of two quarks. Any gluon can be assembled from the colors of a quark and an antiquark or from the colors of three quarks. The procedure is a formal one only; antiquarks and gluons should not be considered physical composites of quarks.

In the absence of gluon charges the strong force might be expected to vary with distance in much the way electromagnetism does. Since the gluons are massless, like the photon, the force would have an infinite range but would decrease in intensity as the square of the distance. The fact that the gluons carry color charges alters the character of the force. Because the cloud of virtual gluons spreads out the color charge, the color force between two quarks does not increase as fast as the electromagnetic force does when the distance between the particles is reduced. As a result the coupling constant of QCD decreases as the distance at which it is measured gets smaller (unlike the coupling constant of QED, which increases at close range). The quarks are said to be asymptotically free, meaning that the QCD coupling constant falls to zero as the distance approaches zero.

Asymptotic freedom was discovered by H. David Politzer, who is now working at the California Institute of Technology, and by David Gross and Frank Wilczek of Princeton University. It has been tested and confirmed in a number of experiments that probe the quark structure of hadrons at small distances. The nature of the strong interactions between quarks at longer distances is not as well established, but it seems the force may not decrease as the square of the distance but may instead remain constant regardless of distance. Unlimited energy would then be needed to separate two color charges, which would explain why quarks seem to be confined permanently in hadrons.

Quantum chromodynamics is called an $SU(3)$ theory, where $SU(3)$ is another term from the theory of groups. The 3 refers to the three colors that are transformed into one another by the gluons. The S indicates that the sum of the color charges in each $SU(3)$ family is zero. In a close analogy with the $U(1)$ of QED, the $SU(3)$ of QCD describes a group of symmetries of the theory that is associated with the conservation of color charge and the masslessness of the gluons.

The higher order of symmetry in the color-$SU(3)$ theory can be suggested in a geometric presentation. The three color charges $R - G$, $G - B$ and $B - R$ can be represented by position along three axes on a

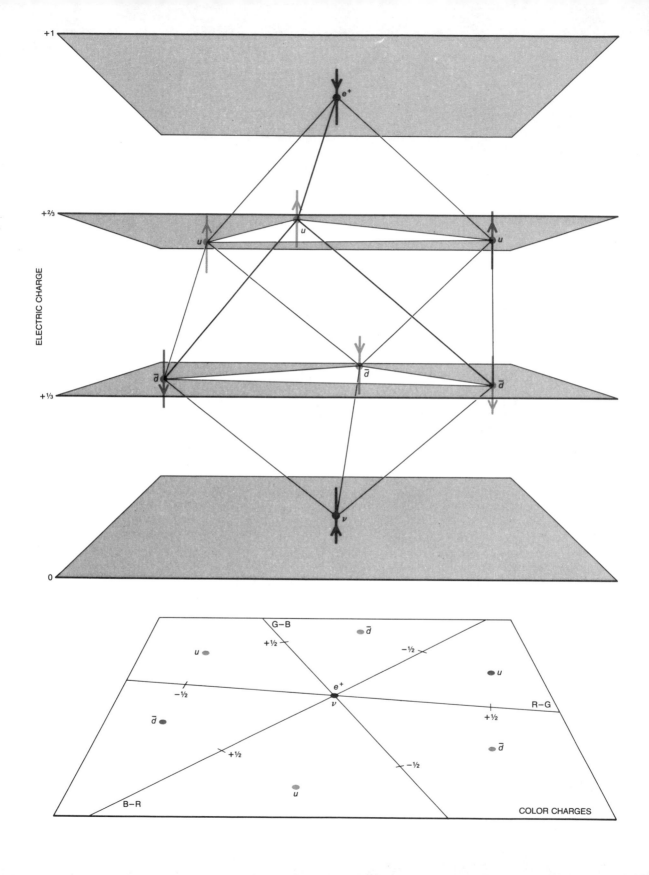

ELECTRIC CHARGE

+1

+⅔

+⅓

0

e⁺

u

u

u

d̄

d̄

d̄

ν

G–B

+½

ū d̄

u

–½

–½

d̄

e⁺

ν

R–G

+½

d̄

+½

–½

u

B–R

COLOR CHARGES

plane. The axes are symmetrically arrayed at angles of 120 degrees to one another. If the three colors are placed according to their component color charges on such a graph, they are found to lie at the vertexes of an equilateral triangle. The anticolors lie opposite the corresponding colors, so that they form another triangle turned 180 degrees from the first one. The two superposed triangles form a Star of David.

A hint of the still greater symmetry in the unified theory is provided by adding a third dimension to the graph (see Figure 29). Suppose the color charges already plotted are assumed to be those of the u quark and the $\bar{d}$ antiquark. Two leptons are now added: the electron-type neutrino (ν) and the positron (e^+). Since the leptons have no color charge, they lie at the origin of the three axes in the center of the plane. The third dimension is electric charge: each particle is to be displaced vertically by an amount proportional to its electric charge. The neutrino remains in its place, but the three $\bar{d}$ antiquarks are shifted upward one-third of a unit, the u quarks move upward two-thirds of a unit and the positron moves upward one unit. If the vertical and horizontal scales are suitably chosen, the eight particles define the vertexes of a cube tipped on one corner. That quarks and leptons can be arranged in the configuration of this simple solid suggests some underlying connection between them.

In order to discuss the last of the three forces, the weak force, it is necessary to introduce another property of the elementary particles: spin angular momentum. All the leptons and all the quarks are observed to have the same fixed quantity of angular momentum, equal to ½ when measured in fundamental units. The particles can be imagined as spinning on an internal axis, like the earth or like a top, but without loss of energy. The angular momentum is represented by a vector, or arrow, along the axis of spin.

A particle with one-half quantum of intrinsic spin

Figure 29 CUBIC SYMMETRY emerges when certain properties of elementary particles are graphed in three dimensions. The position of each particle on the horizontal plane is determined by three kinds of "color charge." Quarks designated u lie at the vertexes of an equilateral triangle; antiquarks labeled $\bar{d}$ form an oppositely oriented triangle. Leptons, represented by the positron (e^+) and the neutrino (ν) lie at the center of the plane. When each particle is displaced vertically by a distance proportional to its electric charge, the cube takes shape.

can have only two possible orientations; in the simplest case, where the particle is in motion, the spin vector can point either in the same direction the particle is moving or in the opposite direction. The two orientations represent two distinguishable states of the particle. The state in which the vector is parallel to the direction of motion is said to be right-handed because when the fingers of the right hand are wrapped around the particle in the same sense as the spin, the thumb indicates the direction of motion. When the spin axis is aligned the opposite way, the thumb of the left hand indicates the direction of motion and so the particle is said to be left-handed (see Figure 30).

In general handedness of a particle can be reversed simply by bringing the particle to rest and accelerating it in the opposite direction without disturbing the spin. Hence most particles necessarily have both left- and right-handed components. The exceptions are massless particles, and the reason they are exceptions is that a massless particle must always move with the speed of light and can never be brought to rest. Therefore the handedness of a massless particle can never change. Among the quarks and leptons the only particles that might be massless are the neutrinos. Experimentally only left-handed neutrinos and right-handed antineutrinos have been observed; right-handed neutrinos and left-handed antineutrinos are presumed not to exist.

The introduction of handedness has the effect of almost doubling the number of distinguishable elementary particles, a number that is already fairly large. In the first generation of particles there are two leptons (the electron and the electron-type neutrino) and two quark flavors (u and d). The three quark colors give a total of eight particles, and accounting for the corresponding eight antiparticles raises the total to 16. If all the particles had been left- and right-handed components, the inclusion of handedness would double the number yet again. Actually, because there is no right-handed neutrino or left-handed antineutrino, the total number of distinct particle and antiparticle states is 30. It is these 30 states that must be accommodated in a unified theory.

States of different handedness must be distinguished because the weak force acts differently on the left-handed and on the right-handed components of a particle. Like the other forces, the weak force is associated with a charge, and the intrinsic strength of the weak interaction can be defined by

RIGHT-HANDED STATE

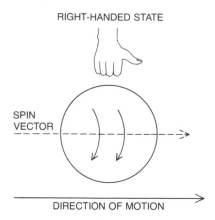

SPIN
VECTOR

DIRECTION OF MOTION

LEFT-HANDED STATE

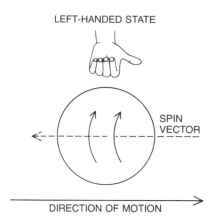

SPIN
VECTOR

DIRECTION OF MOTION

Figure 30 HANDEDNESS OF A PARTICLE. When the spin vector of a particle is parallel to its direction of motion, the particle is said to be right-handed. When the spin vector is antiparallel, the particle is said to be left-handed. Electromagnetism and the strong force are indifferent to handedness but it has an important influence on the weak interactions.

means of a dimensionless coupling constant. The weak charge is unusual, however, in that it is assigned on the basis of handedness. Only left-handed particles and right-handed antiparticles bear a weak charge; the right-handed particles and the left-handed antiparticles are neutral with respect to the weak force and do not participate in weak interactions.

Because the weak charges of the left- and the right-handed electron (for example) are different, the weak charge cannot be conserved. The value of the weak charge depends on which way the electron is moving, and the value must change when the motion changes. The weak charge could be conserved only if the leptons and the quarks were all massless, since in that case none of the particles could come to a stop and reverse direction.

The weak force acts on doublets of particles. The theory that describes it is an $SU(2)$ theory, in which the two members of a doublet can be transformed into each other. For example, the left-handed neutrino and the left-handed electron make up one doublet; they are assigned weak charges of $+\frac{1}{2}$ and $-\frac{1}{2}$ respectively. The left-handed u quark and the left-handed d quark compose a second doublet (or three doublets if each color is counted separately), and they also have respective weak charges of $+\frac{1}{2}$ and $-\frac{1}{2}$. Four right-handed antiparticles form the remaining doublets: the positron, the electron-type antineutrino, the $\bar{d}$ antiquark and the $\bar{u}$ antiquark. Each right-handed antiparticle has a weak charge

opposite to that of the corresponding left-handed particle. Six particles remain to be accounted for: the right-handed components of the electron, the d quark and the u quark and the left-handed components of the positron, the $\bar{d}$ antiquark and the $\bar{u}$ antiquark. They do not form doublets but remain isolated as singlets, and they have a weak charge of zero.

Three particles associated with the weak $SU(2)$ symmetry mediate transitions between the members of each doublet. The intermediary particles are the W^+, with both a weak charge and an electric charge of $+1$; the W^-, with weak and electric charges of -1, and the W^0, which is neutral with respect to both the weak and the electromagnetic forces. The W^0, like the photon and the G_1 and G_2 gluons, conveys a force between particles that carry charge, but it does not alter any of their properties. The W^+ and the W^-, on the other hand, transform the flavors of particles. A left-handed electron can emit a W^- and thereby be converted into a left-handed neutrino; in the process the electric charge changes from -1 to 0 and the weak charge goes from $-\frac{1}{2}$ to $+\frac{1}{2}$. The most familiar weak process is nuclear beta decay, in which a neutron (quark composition udd) emits an electron and an antineutrino and is converted into a proton (uud). Analyzed at a finer scale, the event begins when a d quark emits a virtual W^- and becomes a u quark; the W^- subsequently decays to yield the electron and the antineutrino.

In events such as these one can perceive some tantalizing relations between the weak force and electromagnetism. First, the W particles of the weak force carry weak charge and electric charge in the same amount. Second, in the structure of the weak singlets and doublets there is a curious fixed relation between weak charge and electric charge. The electric charge of a particle is invariably equal to the sum of the particle's weak charge and the average electric charge of the singlet or doublet of which the

	WEAK CHARGE	$U(1)$ CHARGE	ELECTRIC CHARGE	PARTICLES	TRANSITIONS
DOUBLETS	$+\frac{1}{2}$		0	ν_{LEFT}	
		$-\frac{1}{2}$			W^+ W^-
	$-\frac{1}{2}$		-1	e^-_{LEFT}	
	$+\frac{1}{2}$		$+\frac{2}{3}$	u_{LEFT}	
		$+\frac{1}{6}$			W^+ W^-
	$-\frac{1}{2}$		$-\frac{1}{3}$	d_{LEFT}	
	$+\frac{1}{2}$		$+1$	e^+_{RIGHT}	
		$+\frac{1}{2}$			W^+ W^-
	$-\frac{1}{2}$		0	$\bar{\nu}_{\text{RIGHT}}$	
	$+\frac{1}{2}$		$+\frac{1}{3}$	$\bar{d}_{\text{RIGHT}}$	
		$-\frac{1}{6}$			W^+ W^-
	$-\frac{1}{2}$		$-\frac{2}{3}$	$\bar{u}_{\text{RIGHT}}$	
SINGLETS	0	-1	-1	e^-_{RIGHT}	
	0	$+1$	$+1$	e^+_{LEFT}	
	0	$+\frac{2}{3}$	$+\frac{2}{3}$	u_{RIGHT}	
	0	$-\frac{2}{3}$	$-\frac{2}{3}$	$\bar{u}_{\text{LEFT}}$	
	0	$-\frac{1}{3}$	$-\frac{1}{3}$	d_{RIGHT}	
	0	$+\frac{1}{3}$	$+\frac{1}{3}$	$\bar{d}_{\text{LEFT}}$	

Figure 31 WEAK CHARGE depends on the handedness of a particle. Left-handed particles and right-handed antiparticles form doublets in weak interactions and are assigned weak charges of plus or minus ½. The W^+ and W^- trans- form one member of the doublet into another member. Right-handed particles and left-handed antiparticles remain singlets and have no weak charge, so no weak transitions among them are possible.

particle is a member. This average charge I shall designate the $U(1)$ charge. For the singlets the $U(1)$ charge is merely the electric charge of the particle itself and the rule is little more than a tautology: it states that the electric charge is equal to the electric charge, since the weak charge of a singlet particle is always zero. For the weak doublets, however, the relation of the charges is more interesting. It is noteworthy that the relation remains valid both for the lepton doublets, where the charges being averaged are integers, and for the quark doublets, which are made up of fractional charges.

Like the other charges, the $U(1)$ charge is associated with a symmetry. The $U(1)$ symmetry that is extracted from the weak interactions in this way has a particle associated with it, which I shall call V^0. Like the W^0 and the photon, the V^0 has neither electric nor weak charge. Indeed, because the weak, the $U(1)$ and the electric charges are related, the three carrier particles must also be related. It turns out that the W^0 and the V^0 are not observed in nature as pure states but instead appear only as mixtures. One such mixture of the W^0 and the V^0 is the photon; the other possible combination is identified as a particle labeled Z^0. Both the photon and the Z^0 mediate interactions in which a particle is transformed into itself, so that its identity is unchanged. They differ, however, in that the photon couples only particles that have electric charge, whereas the Z^0 couples those that have weak charge, including the neutrinos. By deriving the photon and the W and Z particles from the same theory, the weak force and electromagnetism are partially unified. The combined theory is designated by the product of the groups it incorporates: $SU(2) \times U(1)$ (see Figure 32).

By analogy with the $U(1)$ of quantum electrodynamics and the $SU(3)$ of quantum chromodynamics, one might expect the $SU(2)$ and the $U(1)$ of the weak interaction to be exact symmetries of the theory. The weak charge would then be exactly conserved and the W and Z particles would be massless and of infinite range. As noted above, however, weak charge is not invariably conserved; furthermore, the weak force is observed to have an exceedingly short range of perhaps 10^{-15} centimeter. The reason is that the W^+, the W^- and the Z^0 have large masses, approaching 100 GeV, or 100 times the mass of the proton. What becomes of the $SU(2) \times U(1)$ symmetry of the weak and the electromagnetic interactions under these conditions? Moreover, if the photon and the W and Z particles are closely related,

how can the one particle remain massless if the others acquire a large mass?

Answering these questions was an essential step in formulating the combined theory of weak and electromagnetic forces. The answer that is now favored is that the underlying force is indeed symmetrical, and in some hypothetical initial state all the carriers of the weak force would be massless. What is not symmetrical is the quantum-mechanical vacuum, the milieu in which the force operates. The structure of the vacuum spontaneously breaks the $SU(2) \times U(1)$ symmetry, giving a mass to the three carriers of the weak force but not to the photon. An analogous loss of symmetry can be observed in a crystal of salt. The ions in the crystal do not favor any one direction in space over any other; the ions are rotationally symmetrical. In the lattice structure of the crystal, however, the symmetry is broken, and certain directions, such as those parallel to lattice lines, have a special status.

The same analogy can illuminate another characteristic of a spontaneously broken symmetry: The appearance of the world depends on the scale at which it is examined. In the crystal three scales of distance can be distinguished. At distances of much less than 10^{-8} centimeter (the size of an atom) one sees the internal structure of the atom, which is fully symmetrical and unaffected by the organization of the crystal. At distances of about 10^{-8} centimeter the forces responsible for the cohesion of the atoms in the crystal become important, and the phenomena observed are extremely complex. At distances much greater than 10^{-8} centimeter the geometry of the crystal becomes fully apparent and the rotational symmetry of the atoms is conspicuously broken.

A similar hierarchy of distance scales can be defined in connection with the spontaneous breakdown of the $SU(2) \times U(1)$ symmetry, but the critical distance is smaller: about 10^{-16} centimeter. At distances much smaller than this the full symmetry is expressed. At such close range the massive W and Z particles are exchanged as readily as massless photons, and so the weak and the electromagnetic forces are effectively unified. Another way of expressing the same idea is to point out a relation between distance and energy. According to the uncertainty principle, the energy needed to probe or resolve a given distance is inversely proportional to the distance. An experiment that examined the

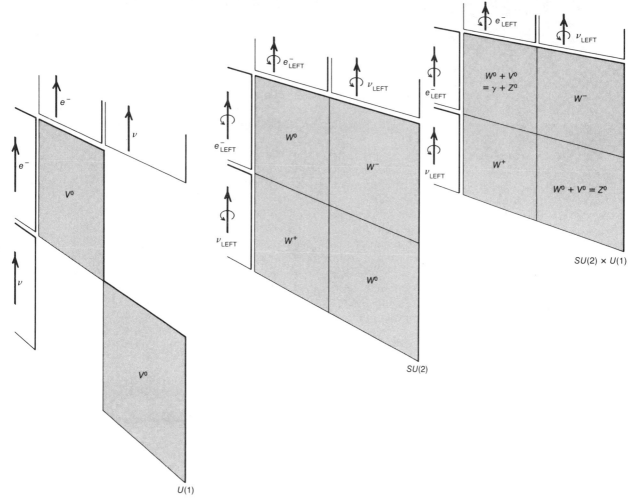

Figure 32 WEAK FORCE AND ELECTROMAGNETISM can be described by the $SU(2) \times U(1)$ theory. The $SU(2)$ part induces all possible transformation of the two objects, represented here by the left-handed components of the electron and the neutrino. The $U(1)$ part is associated with the V^0, which is capable only of identity transformations. In the combined theory the W^0 and V^0 become mixed; the observed combinations are the photon (γ) and a carrier of the weak force called Z^0.

structure of a particle at a range of less than 10^{-16} centimeter would have to be done at an energy of more than 100 GeV. At this energy W and Z particles can be freely created, and the mass difference between them and the photon is negligible; all the intermediary particles are light compared with the energy of the experiment.

At a distance of about 10^{-16} centimeter the complex phenomena responsible for breaking the $SU(2) \times U(1)$ symmetry begin to intrude. W and Z particles are still observed, but they look quite different from the photon because their masses are comparable to the energy of the experiment. At still larger distances the symmetry between the photon and the W and Z particles is completely obscured; indeed, there is insufficient energy to create a real W and Z, and so they cannot be observed directly. Only the rare, short-range effects of their virtual exchange (such as the beta decay of the neutron) can be detected. This is the present domain of particle physics.

The concept of spontaneous symmetry breaking resolves the question of weak-charge conservation. At an energy much greater than 100 GeV, where the $SU(2) \times U(1)$ symmetry is observed directly, the mass of a quark or a lepton is negligible; the handedness of the particles is therefore essentially fixed, and so the weak charge is effectively conserved. At low energy, where the symmetry is spontaneously broken, the weak charge is not conserved but can disappear into the vacuum when a massive particle changes handedness.

This theory of the weak and electromagnetic interactions was worked out in the 1960's by Sheldon Lee Glashow and by Steven Weinberg of Harvard University and by Abdus Salam of the International Center for Theoretical Physics in Trieste. Glashow was the first to deduce the form of the $SU(2) \times U(1)$ theory, but he did not know how to incorporate into it the masses of the W and Z particles. Weinberg and Salam separately found the $SU(2) \times U(1)$ form later and applied the notion of spontaneous symmetry breaking, thereby formulating a consistent theory.

The $SU(2) \times U(1)$ theory is only a partial unification because it still includes two distinct forces, each with its own symmetry group and its own coupling constant. The ratio of the coupling constants is a free parameter, which must be chosen to fit experimental results. Another deficiency of the theory is that electric charge is only partly quantized. The construction of doublets of particles related by $SU(2)$ transformations requires that all differences between electric charges be integers, so that each particle can change its identity by emitting a W^+ or a W^-. The average charge of the doublets, however, is not quantized. The average charge is the $U(1)$ charge, which is defined in conjunction with the electromagnetic charge of quantum electrodynamics; as in that theory there is no fundamental reason for confining the charges to integer values. Indeed, in the doublets made up of quarks the interval between the charges is an integer, but the actual charges are fractional.

The endeavor to construct a unified theory of the weak, the strong and the electromagnetic forces should not be seen as an attempt to do away with the color $SU(3)$ model or the $SU(2) \times U(1)$ model. The theories of the individual forces work too well for them to be discarded. What a unifying theory can provide is a superstructure in which the $SU(3)$ and $SU(2) \times U(1)$ theories can be embedded. The superstructure would take the form of a larger symmetry in which quarks and leptons would be closely related.

The search for such a larger symmetry must begin with the search for a larger group, one that includes both $SU(3)$ and $SU(2) \times U(1)$ as component structures. Many groups have this property, but one can-

	ELECTRIC CHARGE	WEAK CHARGE	R–G CHARGE	G–B CHARGE
d_{RIGHT}^{RED}	$-\frac{1}{3}$	0	$+\frac{1}{2}$	0
d_{RIGHT}^{GREEN}	$-\frac{1}{3}$	0	$-\frac{1}{2}$	$+\frac{1}{2}$
d_{RIGHT}^{BLUE}	$-\frac{1}{3}$	0	0	$-\frac{1}{2}$
e_{RIGHT}^{+}	$+1$	$+\frac{1}{2}$	0	0
$\bar{\nu}_{RIGHT}$	0	$-\frac{1}{2}$	0	0
	$= 0$	$= 0$	$= 0$	$= 0$

Figure 33 INTEGRATED FAMILY of five elementary particles, each assigned values of electric charge, of weak charge and of two color charges. For each kind of charge the sum of the values assigned to all the particles is zero. Furthermore the particles that bring about transformations within the family carry each kind of charge. It follows that all the charges of the particles are quantized.

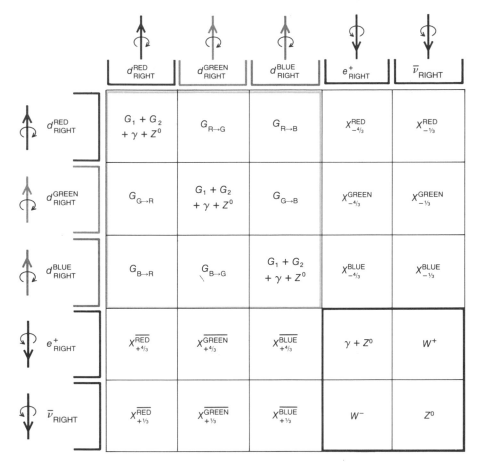

Figure 34 *SU*(5) **SYMMETRY encompasses all possible transitions between particles in the integrated family of five. The *SU*(3) symmetry of the strong force is embodied in the three-by-three matrix (*upper left*); the *SU*(2) symmetry of the weak force appears in the two-by-two matrix (*lower right*). The *U*(1) symmetry appears in the couplings of the photon (γ) and the *Z*⁰ along the diagonal. The *SU*(5) theory postulates 12 new intermediary particles, labeled *X*.**

didate has numerous advantages that distinguish it from all the others. This group is *SU*(5), the group of all possible transformations of five distinct objects, or of a five-by-five matrix. It is the smallest simple group that can accommodate the constituent *SU*(3) and *SU*(2) × *U*(1) symmetries, and I believe it may be the full symmetry group of nature. A unified theory based on *SU*(5) symmetry was worked out in 1973 by Glashow and me.

In the simplest representation of the *SU*(5) group the five objects are the right-handed components of the *d* quark in each of the colors red, green and blue, the right-handed component of the positron and the right-handed component of the electron-type antineutrino (which has only a right-handed component). Each of the five particles is assigned a value

for each of four independent charges: electric charge, weak charge and two color charges, which I shall take to be R − G and G − B (see Figure 33).

Twenty-four intermediary particles provide for all possible transitions between these five states of matter (see Figure 34). Four of the particles are the photon, the *Z*⁰ and the gluons G_1 and G_2, which are directly associated with the four fundamental charges. Since these particles carry no charge, they take part only in those interactions where the identity of a particle does not change. Of the remaining 20 intermediary particles eight are already familiar. They are the *W*⁺ and *W*⁻, which can convert a positron into an antineutrino and vice versa, and the six gluons that transform the colors of the quarks. With this complement of 12 carrier particles all interac-

| | KIND OF CHARGE | | | |
	ELECTRIC	WEAK	R–G	G–B
d_{RIGHT}^{RED}	−1/3	0	+1/2	0
+ e_{RIGHT}^{+}	+1	+1/2	0	0
u_{LEFT}^{RED}	+2/3	+1/2	+1/2	0
d_{RIGHT}^{GREEN}	−1/3	0	−1/2	+1/2
+ e_{RIGHT}^{+}	+1	+1/2	0	0
u_{LEFT}^{GREEN}	+2/3	+1/2	−1/2	+1/2
d_{RIGHT}^{BLUE}	−1/3	0	0	−1/2
+ e_{RIGHT}^{+}	+1	+1/2	0	0
u_{LEFT}^{BLUE}	+2/3	+1/2	0	−1/2
d_{RIGHT}^{RED}	−1/3	0	+1/2	0
+ $\bar{\nu}_{RIGHT}$	0	−1/2	0	0
d_{LEFT}^{RED}	−1/3	−1/2	+1/2	0
d_{RIGHT}^{GREEN}	−1/3	0	−1/2	+1/2
+ $\bar{\nu}_{RIGHT}$	0	−1/2	0	0
d_{LEFT}^{GREEN}	−1/3	−1/2	−1/2	+1/2

Figure 35 FAMILY OF 10 PARTICLES is assembled by forming all possible pairs of the five states that make up the simplest family of $SU(5)$.

tions observed in nature up to now can be accounted for. The $SU(5)$ group includes 12 more intermediary particles, however, which are needed if the theory is to have the maximum possible symmetry. The 12 extra particles are labeled X, and they mediate the interconversion of leptons and quarks. Each X particle carries weak charge, color charge and electric charge; the electric charges have values of plus or minus 1/3 and plus or minus 4/3.

As with the distribution of color charges in $SU(3)$, the table of charge assignments in the $SU(5)$ theory has some intriguing regularities. For each kind of charge the sum of the charges assigned to the five particles is zero. For example, each of the three quark colors has an electric charge of −1/3, but these charges are balanced by the positron's electric charge of +1. A related observation is that all four varieties of charge are carried by at least some of the $SU(5)$ intermediary particles. The gluons have color, the W^+ and W^- have both weak charge and electric charge and the X particles carry all four forms of charge.

From these two facts it can be deduced that all the charges are necessarily quantized. All electric charges must be multiples of 1/3; if a particle with some different charge were accepted into the fam-

| | | KIND OF CHARGE | | | |
		ELECTRIC	WEAK	R–G	G–B
	d^{BLUE}_{RIGHT}	−1/3	0	0	−1/2
+	$\bar{\nu}_{RIGHT}$	0	−1/2	0	0
	d^{BLUE}_{LEFT}	−1/3	−1/2	0	−1/2
	e^+_{RIGHT}	+1	+1/2	0	0
+	$\bar{\nu}_{RIGHT}$	0	−1/2	0	0
	e^+_{LEFT}	+1	0	0	0
	d^{GREEN}_{RIGHT}	−1/3	0	−1/2	+1/2
+	d^{BLUE}_{RIGHT}	−1/3	0	0	−1/2
	$\bar{u}^{RED}_{LEFT}$	−2/3	0	−1/2	0
	d^{RED}_{RIGHT}	−1/3	0	+1/2	0
+	d^{BLUE}_{RIGHT}	−1/3	0	0	−1/2
	$\bar{u}^{GREEN}_{LEFT}$	−2/3	0	+1/2	−1/2
	d^{RED}_{RIGHT}	−1/3	0	+1/2	0
+	d^{GREEN}_{RIGHT}	−1/3	0	−1/2	+1/2
	$\bar{u}^{BLUE}_{LEFT}$	−2/3	0	0	+1/2

Figure 35 (continued).

ily, the $SU(5)$ carrier particles could not be emitted or absorbed by it without violating the conservation of charge. Moreover, it is not just the minimum interval between charges that is fixed; the actual values of the charges are determined by the requirement that the total charge be zero. Here at last is an explanation of the quantization of electric charge. The same requirement explains the exact commensurability of the lepton and quark charges, which in turn implies the exact neutrality of the atom. In addition the intriguing coincidence that all color-neutral systems of particles have integral electric charge follows from the organization of the family.

What about the remaining particles of the first generation? One of the most elegant features of the $SU(5)$ theory is that the five right-handed particles in the smallest family of $SU(5)$ can be combined in pairs to yield a family of 10 left-handed particles. These 10 states make up the next-simplest representation of the group. They are the left-handed components of the d quark, the u quark and the $\bar{u}$ antiquark (in three colors each) and of the positron. As in the assembly of antiquarks and gluons from the basic triplet of quark colors, this process should not be interpreted as a physical prescription for building the particles. A left-handed lepton or quark is not actually composed of two right-handed particles in a bound state. Nevertheless, the 10 possible ways of forming pairs of the five right-handed states yield all the correct charges for the left-handed particles.

The transitions available to any particle are also correctly given by this scheme of composition by pairs. They include both quark-lepton and quark-antiquark transitions. What is equally important, transitions that are not observed are not allowed by the structure of the group. Each family of particles is closed, or complete in itself. Every allowed transition gives rise to another particle in the same family and no other transitions are possible.

The family of five right-handed states and the family of 10 left-handed states hold a total of 15 particles. Two more families of slightly different form can be constructed to accommodate the remaining 10 right-handed particles and five left-handed particles, which are the antiparticles of the 15 states in the first two families. Hence all 30 elementary states of the first generation have a place in the theory, and there are no empty places (see Figure 35). Equivalent representations of the higher generations can be constructed by substituting the muon or the tau lepton for the electron, the s or b quark for the d quark, and so on.

Let me review what has been done so far. First the group $SU(5)$ was chosen as the smallest group in which $SU(3)$ and $SU(2) \times U(1)$ could be embedded. Next five right-handed components of particles were selected as members of the simplest $SU(5)$ family. The remaining components then had to fit into some other family of $SU(5)$, and they did. With nothing left out and nothing left over, they fit into the next-simplest family. Moreover, the composition of the derived family was specified by a simple procedure for combining particles in pairs. It is important to note that this assembly procedure did not have to work. In many groups other than $SU(5)$ it would not work. It represents the first, the simplest and in some respects the most remarkable success of the $SU(5)$ theory.

The most obvious significance of the $SU(5)$ unification is that the leptons and the quarks are no longer irreconcilably different. Instead they are members of a single family, and a quark can be converted into a lepton (or vice versa) as easily as one quark can be converted into a different quark or one lepton into a different lepton. A further consequence of the unification is that the weak, the strong and the electromagnetic forces should all have the same strength, or the same coupling constant. Neither of these expectations is satisfied in the world as it appears today. In the millions of elementary-particle interactions recorded by physicists not a single instance of quark-lepton conversion has been observed. Furthermore, the coupling constants of the three forces differ by large factors: the strong force is roughly 100 times as strong as electromagnetism, with the weak force somewhere in between. Thus if $SU(5)$ is a symmetry of nature, it is evidently a badly broken one.

The symmetry could be broken by a mechanism similar to the one that breaks the $SU(2) \times U(1)$ symmetry of the weak and electromagnetic forces. In this way the X particles would be given a large mass and the effects of X-particle exchange would be strongly suppressed. In $SU(5)$, however, the breakdown would have to come at a much higher energy or equivalently at a much smaller distance than in $SU(2) \times U(1)$. This distance is the unification scale, the distance at which the full symmetry of the theory becomes manifest.

With the $SU(5)$ theory one can speculate about what the world would look like at various scales of distance or energy. In an experiment that probed distances much smaller than the unification scale the $SU(5)$ gauge invariance of the world would be readily apparent. All interactions, including the quark-lepton and quark-antiquark transformations, would be on an equal footing because all the $SU(5)$ intermediary particles (the photon, the gluons, the W and Z particles and the X particles) would be created with essentially equal probability. The masses of the W, Z and X particles would hardly distinguish them from the photon and gluons because the masses would be small compared with the energy of the experiment.

At distances close to the unification scale the complicated physics associated with the spontaneous breakdown of $SU(5)$ symmetry would be observed. X particles would be emitted, but their mass would make them quite different from all the other particles. At distances much greater than the unification scale (but still much smaller than 10^{-16} centimeter) the $SU(5)$ symmetry would be almost completely hidden. The X particles could no longer be created as real particles, and so the leptons and the quarks would become segregated in separate families, with little communication between them. On the other hand, the $SU(2) \times U(1)$ symmetry would remain intact, so that little distinction could be made between the weak and the electromagnetic interactions. At distances greater than 10^{-16} centimeter the $SU(2) \times U(1)$ symmetry would also be broken and there would be three distinct forces.

Spontaneous symmetry breaking can also explain the disparities among the coupling constants. The crucial element in the explanation is the effect of the virtual particles in the vacuum surrounding a point charge. As noted earlier, if a positive electric charge is surrounded by virtual photons and virtual electron-positron pairs (see Figure 36), the photons have little effect, but the charged virtual particles are polarized, so that the negative virtual charges cluster around the positive real charge, reducing its effective magnitude. If a positive color charge is enveloped in a cloud of virtual gluons and virtual quark-antiquark pairs, the quarks and antiquarks become polarized much as electrons and positrons do, but the gluons act differently from photons. Whereas the photon is electrically neutral, some of the gluons have a color charge, and it is predominantly of the same polarity as the real charge. The cloud of virtual gluons that envelops a quark effectively spreads out the color charge, with the result

that the coupling constant decreases when it is measured at closer range. The ultimate result is that the electromagnetic interaction becomes larger at close range, whereas the strong interaction becomes smaller. Virtual W particles have a similar effect on a weak charge, although the effect is somewhat smaller because there are fewer W particles that carry the weak charge than there are gluons that carry the color charge. In the $U(1)$ theory, on the other hand, the fact that the Z^0 has no charge gives rise to a quite different phenomenon. Because of the polarization of virtual electrons and positrons the $U(1)$ coupling constant increases at close range.

A simple inference can be made from these trends in the coupling constants. At long range the $SU(3)$ coupling constant of the strong force is the largest, but it also decreases the fastest; the $SU(2)$ constant of the weak force is smaller and decreases slower; the $U(1)$ constant is the smallest, but it increases as the distance is reduced. Hence it appears there may be some distance at which all three coupling constants have approximately the same value.

At distances of less than 10^{-16} centimeter all the coupling constants are fairly small and the way they vary with range or energy can be calculated. Values

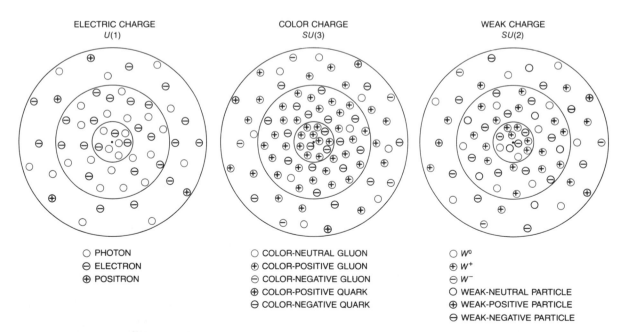

ELECTRIC CHARGE
$U(1)$

COLOR CHARGE
$SU(3)$

WEAK CHARGE
$SU(2)$

○ PHOTON
⊖ ELECTRON
⊕ POSITRON

○ COLOR-NEUTRAL GLUON
⊕ COLOR-POSITIVE GLUON
⊖ COLOR-NEGATIVE GLUON
⊕ COLOR-POSITIVE QUARK
⊖ COLOR-NEGATIVE QUARK

○ W^0
⊕ W^+
⊖ W^-
○ WEAK-NEUTRAL PARTICLE
⊕ WEAK-POSITIVE PARTICLE
⊖ WEAK-NEGATIVE PARTICLE

Figure 36 CLOUD OF VIRTUAL PARTICLES surrounds a central point charge and alters its response to the force. For electric charge (*left*) the cloud reduces the effective magnitude; for color charge (*center*) force is reduced inside the cloud; for weak charge (*right*) the force also becomes feebler at small distances.

can be determined at progressively smaller distances; the distance at which the three constants converge is the unification scale. If two of the constants are followed in this way, the value of the third constant can be predicted at any energy. Such calculations have been done using as input the strong coupling constant and the electromagnetic coupling constant; the latter, again, is a combination of the $U(1)$ and $SU(2)$ coupling constants. The results yield values of the unification scale and of the ratio of the $U(1)$ and $SU(2)$ constants, which had been an arbitrary parameter in nonunified theories.

In 1974 I made such a calculation with Helen R. Quinn (who is now at the Stanford Linear Accelerator Center) and Weinberg for a class of unified theories that includes $SU(5)$. We obtained a distance of about 10^{-29} centimeter for the unification scale and a value of .2 for the $U(1)$-to-$SU(2)$ ratio. At the time the results were not encouraging because measurements of the ratio suggested a value of about .35. Since then refined measurements of the coupling-constant ratio have given lower values. The ratio is now thought to be equal to .22 plus or minus .02, in agreement with the theoretical result.

Another prediction of the $SU(5)$ model that can be tested at accessible energy was worked out in 1977 by Andrzej Buras, John Ellis, Mary K. Gaillard and Demetres V. Nanopoulos of the European Organization for Nuclear Research (CERN) in Geneva. They found that in the simplest version of the $SU(5)$ theory the ratio of the mass of the b quark to the mass of the tau lepton can be determined. As in the case of the coupling constants, the masses are expected to be equal at the unification scale, but at greater distances the b quark is heavier because of its color charge. The mass ratio at low energy was calculated to be about $3:1$. The mass of the tau is known: it is a little less than twice the mass of the proton. The mass of the b quark is not as certain because the quark cannot be examined in isolation. The best estimate at present is about five times the photon mass, making the ratio of the masses $5:2$.

The unification scale of 10^{-29} centimeter is an extraordinarily small distance. (If a single proton were inflated to the size of the sun, the unification scale would still be less than a micrometer.) Implicit in the $SU(5)$ unification is the assumption that no new physical principles will be encountered in the entire range of distances between 10^{-16} and 10^{-29} centimeter; in particular, it must be assumed that the way the coupling constants vary with distance remains unchanged (see Figure 37). Such an assumption is of course disturbing, but it is not entirely implausible. There is already a small scale of distance at which new phenomena are expected. At about 10^{-33} centimeter gravitation may become as strong as the other forces, and so any theory that describes events at this scale must include gravitation. I find it encouraging that although the unification scale is small, it is still 10,000 times greater than 10^{-33} centimeter.

A distance of 10^{-29} centimeter corresponds to an energy of about 10^{15} GeV, or roughly 10^{15} times the mass of the proton. The X particles must have a mass equivalent to this energy. For the sake of comparison, the heaviest particles that can now be created with particle accelerators have a mass of about 100 GeV. In order to create X particles the energy would have to be further increased by 13 orders of magnitude, which seems unlikely ever to be feasible.

Even if it is never possible to exhibit a real X particle in the laboratory, the existence of such particles might be demonstrated by detecting events in which a virtual X particle is exchanged. Such exchanges would themselves be extremely rare, since they could take place only when two elementary particles happened to stray within less than 10^{-29} centimeter of each other. Even in the welter of commoner events, however, the exchange of an X particle would be readily discerned because X particles can do something no other particles can do: they can transform a quark into a lepton or a quark into an antiquark. This process puts in question the very stability of matter.

The interactions mediated by X particles differ from all other interactions in that they violate the conservation of a quantity called baryon number. The baryon number of any particle can be defined as one-third of the number of quarks minus one-third of the number of antiquarks. Hence the proton and all other particles made up of three quarks have a baryon number of $+1$, whereas the pi meson and other particles that consist of a quark and an antiquark have a baryon number of 0. Of course the leptons also have a baryon number of 0, since they include no quarks or antiquarks. In the strong, the weak and the electromagnetic interactions the total baryon number can never change. If the conservation of baryon number were an absolute law of nature, the proton could never decay because the proton is the lightest particle with non-

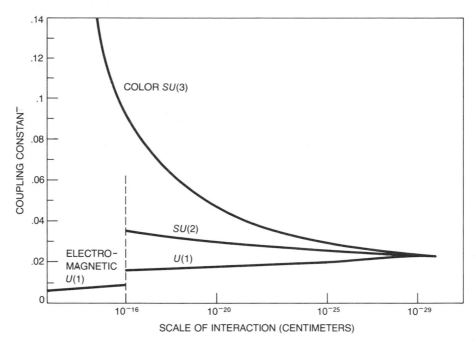

Figure 37 UNIFICATION OF FORCES in the *SU*(5) theory. The intrinsic strength of each force is measured by a dimensionless coupling constant associated with the underlying symmetry. Because of the effects of virtual particles surrounding a charge, the strong *SU*(3) interaction declines in strength as the separation between particles is reduced. The *SU*(2) interaction also becomes weaker at close range, but at a lesser rate. Extrapolation from comparatively long range suggests that the curves converge at a distance of about 10^{-29} centimeter.

zero baryon number. The *SU*(5) unification predicts just such a decay (see Chapter 6, Figure 54).

A possible failure of baryon-number conservation can be illustrated in a proton that forms the nucleus of a hydrogen atom. The proton consists of *u, u* and *d* quarks, with one quark in each of the three colors. If two of the quarks happen to approach to within 10^{-29} centimeter, an X particle can pass between them. For example, a right-handed red *d* quark can emit an X with an electric charge of $-\frac{4}{3}$ and color charges corresponding to the color red. The *d* quark, having lost its color charge and having changed its electric charge from $-\frac{1}{3}$ to $+1$, would thereby become a positron. Meanwhile the X particle could be absorbed by a left-handed green *u* quark, which would be converted into a left-handed $\bar{u}$ antiquark with the color antiblue. The new $\bar{u}$ would combine with the remaining *u* quark to form a neutral pi meson. The baryon numbers of both the positron and the pi meson are zero, so that the total baryon number changes from $+1$ to 0.

If this event were observed in the laboratory, the

exchange of the X particle could not be directly perceived. All that would be seen would be the decay of a proton into a positron and a neutral pi meson. Moreover, that would not be the end of the sequence of events. The positron would subsequently encounter an electron (perhaps the electron of the hydrogen atom) and they would annihilate each other, giving rise to gamma rays, or high-energy photons. The *u* quark and $\bar{u}$ antiquark of the neutral pi meson would eventually annihilate each other in a similar way, releasing more gamma rays. The end result is that a hydrogen atom decays into a state of pure radiation. This process represents a conversion of matter into energy far more efficient than that of nuclear fission or thermonuclear fusion. The fusion of hydrogen atoms to form helium releases less than 1 percent of their mass as energy, whereas the process outlined here liberates 100 percent of the mass.

The abrupt disappearance of a proton and hence of an atom is an event that must happen very rarely or it would have been noticed long ago. Indeed, a low rate is to be expected because particles seldom come within the range where X-particle exchange is possible. Quinn, Weinberg and I employed our calculation of the unification scale to estimate the decay rate of the proton. Our estimate has since been refined by many others, including Buras, Ellis, Gaillard and Nanopoulos of CERN, Terrence J. Goldman and Douglas A. Ross of Cal Tech and William J. Marciano of Rockefeller University. The present estimate is that the average lifetime of the proton is about 10^{31} years.

As a test of the $SU(5)$ unification it is obviously impractical to wait 10^{31} years for a given proton to decay. The age of the universe since the big bang is only about 10^{10} years. The search for proton decay is not hopeless, however. An average lifetime of 10^{31} years implies that in a collection of 10^{31} protons a decay ought to be observed once a year. In 1,000 tons of matter there are about 5×10^{32} protons and neutrons, so that roughly 50 of them can be expected to decay each year. Hence the strategy for detecting events that violate the conservation of baryon number is to monitor everything that goes on in at least 1,000 tons of matter for several years and to distinguish decaying protons and neutrons from commoner events.

Several groups of investigators are planning experiments on this scale. The experiments will be done deep underground or underwater in order to reduce to a minimum the number of cosmic rays passing through the sample of matter. The cosmic rays can give rise to interactions that might be mistaken for proton decay. One experiment is to be installed in a salt mine near Cleveland, another in a silver mine in Utah and a third in an iron mine in Minnesota. Experiments of a slightly smaller scale are planned for two tunnels under the Alps, and smaller experiments are already under way in gold mines in South Dakota and India.

The energy needed to create real X particles may be forever beyond the capabilities of manmade machines, and indeed there may be no process anywhere in the universe today that can generate such a high energy. In an earlier era, however, X particles may have been commonplace. About 10^{-40} second after the big bang the size of the universe was comparable to the unification scale. The universe was then so hot (about 10^{28} degrees Kelvin) that all particles had energies comparable to the X mass. Consequently the $SU(5)$ symmetry was just beginning to break down, and quark-lepton conversions were as frequent as any other interactions. No fundamental distinction could be made between quarks and leptons or between the strong, weak and electromagnetic forces: there was one kind of matter and one force.

A remnant of that era of manifest symmetry may persist in the universe today; in a sense the universe today is the remnant. An old puzzle of astrophysics asks why the universe is built out of matter rather than antimatter. Actually, perhaps the simplest expectation is that there would be equal quantities of matter and antimatter, which would ultimately annihilate each other to leave a universe consisting of nothing but radiation. The $SU(5)$ unification offers a conjectural explanation for the apparent predominance of matter. it is possible that the free interchange of X particles in a brief period after the $SU(5)$ symmetry was broken created more quarks than antiquarks and hence more baryons than antibaryons.

The intriguing speculation that processes violating baryon number might be responsible for the excess of baryons was first made by the Russian physicist Andre Sakharov in 1967. More recently Motohiko Yoshimura of Tohoku University suggested that the baryon-number violation predicted by unified theories has the right properties to account for the observed excess. The idea has since been elaborated by many others, including Ellis, Gaillard, Nanopoulos, Weinberg, Savas Dimopoulos and Leonard Suskind of Stanford University and Sam B. Treiman, Anthony Zee and Wilczek of Princeton. They have shown that an excess of baryons can arise only if the processes that violate the conservation of baryon number look different when they are run backward in time. This condition is satisfied by the $SU(5)$ theory. Thus one item of evidence for the $SU(5)$ unification, albeit indirect and circumstantial evidence, is the very existence of matter.

POSTSCRIPT

My article appeared eight years after the work on which it was based. By then grand unification, of which I describe $SU(5)$ theory as the simplest example, was already a mature idea. There has been little theoretical progress since, but two related areas, the search for proton decay and attempts at unification with gravity, have seen major changes.

The most important change for grand unification has been experimental. As J. M. LoSecco, Fredrick Reines and Daniel Sinclair discuss in Chapter 7, "The Search for Proton Decay," the lifetime of the proton is now known to be greater than predicted by the simplest $SU(5)$ theory. However, this version may still not be ruled out. We still do not understand the mathematics of QCD well enough to be entirely confident in its application, but it does seem likely that the simple version of $SU(5)$ has not been implemented by nature. This should not bother us very much. This version of the theory was based on the extremely strong assumption that there are no heavy particles remaining to be discovered with masses between 100 times the proton mass and a unification scale of some 10^{15} times the proton mass. It was worthwhile to extrapolate the physics that we see in this way, but we should not be too disappointed to discover that nature has done something more interesting.

What is disappointing about the failure of the naive version of the $SU(5)$ model is that it is the only one in which we can make a definite prediction about proton decay. A different version of $SU(5)$ or a closely related model could be right, but it would have many parameters that we cannot determine from the study of physics at currently accessible energies. Thus, in the absence of proton decay we cannot tell if some minor modification of $SU(5)$ is required or if the whole idea of grand unification is wrong, or something in between.

The theoretical attempts to go beyond $SU(5)$ grand unification have been considered by people such as Michael Green [see "Superstrings," SCIENTIFIC AMERICAN, September, 1986] and Haim Harari [see "The Structure of Quarks and Leptons," SCIENTIFIC AMERICAN, April, 1983]. These theoretical ideas are very different from each other, and each is very different from the idea of grand unification.

Harari describes an ambitious attempt to find another level of dynamics to explain some of the properties of the quarks and leptons. This is a wonderful enterprise, but it has not progressed very far because we have no experimental evidence of such substructure. Without it, there is a nearly infinite number of possibilities. Nevertheless, serious theoretical studies in this area are important as eventually machines will be built with which we can see inside the quarks and leptons, if they have any inside. By studying the issues involved we might know what to expect. Green considers an attempt to write an "ultimate" theory, which describes the way the world works down to infinitely short distances using superstrings. This motivation is entirely different from that which led to grand unification.

The $SU(5)$ theory is really very conservative. It builds on what we know from the distances we can study directly without changing the underlying rules. Indeed, the simplest $SU(5)$ theory may be too conservative, and we probably have not seen proton decay because there is more physics on the way to unification. Superstrings, on the other hand, are motivated primarily by considerations of mathematical elegance.

My own guess is that superstrings will remain mere mathematics. I very much hope, however, that we will find some evidence of substructure. This would give us a new and interesting set of mysteries to explore. Meanwhile experimental physicists are pushing the search for proton decay further so that if the proton is found to decay, we will have some direct experimental information about physics at the unification scale. If the idea of grand unification remains purely theoretical, it is not very interesting.

Gauge Theories of the Forces between Elementary Particles

All the basic forces of nature are now described by theories of this kind. The properties of the forces are deduced from symmetries or regularities apparent in the laws of physics.

. . .

Gerard 't Hooft

An understanding of how the world is put together requires a theory of how the elementary particles of matter interact with one another. Equivalently, it requires a theory of the basic forces of nature. Four such forces have been identified, and until recently a different kind of theory was needed for each of them. Two of the forces, gravitation and electromagnetism, have an unlimited range; largely for this reason they are familiar to everyone. They can be felt directly as agencies that push or pull. The remaining forces, which are called simply the weak force and the strong force, cannot be perceived directly because their influence extends only over a short range, no larger than the radius of an atomic nucleus. The strong force binds together the protons and the neutrons in the nucleus, and in another context it binds together the particles called quarks that are thought to be the constituents of protons and neutrons. The weak force is mainly responsible for the decay of certain particles.

A long-standing ambition of physicists is to construct a single master theory that would incorporate all the known forces. One imagines that such a theory would reveal some deep connection between the various forces while accounting for their apparent diversity. Such a unification has not yet been attained, but in recent years some progress may have been made. The weak force and electromagnetism can now be understood in the context of a single theory. Although the two forces remain distinct, in the theory they become mathematically intertwined. What may ultimately prove more important, all four forces are now described by means of theories that have the same general form. Thus if physicists have yet to find a single key that fits all the known locks, at least all the needed keys can be

Figure 38 FOUR BASIC FORCES: strong, weak, electromagnetic and gravitational. As the strong force is observed acting between hadrons, such as the proton and neutron (*solid colored line*), it has a finite range of 10^{-13} cm. The strong force also binds together the quarks that make up hadrons, and it could be expected to follow an inverse-square law (*broken colored line*). Actually the force remains constant regardless of distance (*dotted colored line*). In quantum field theories (*diagrams at left*) the force between two particles is made manifest through the exchange of a third, or virtual, particle, from whose mass the force is determined. Massless virtual particles, such as the photon and graviton, give rise to forces that have infinite range.

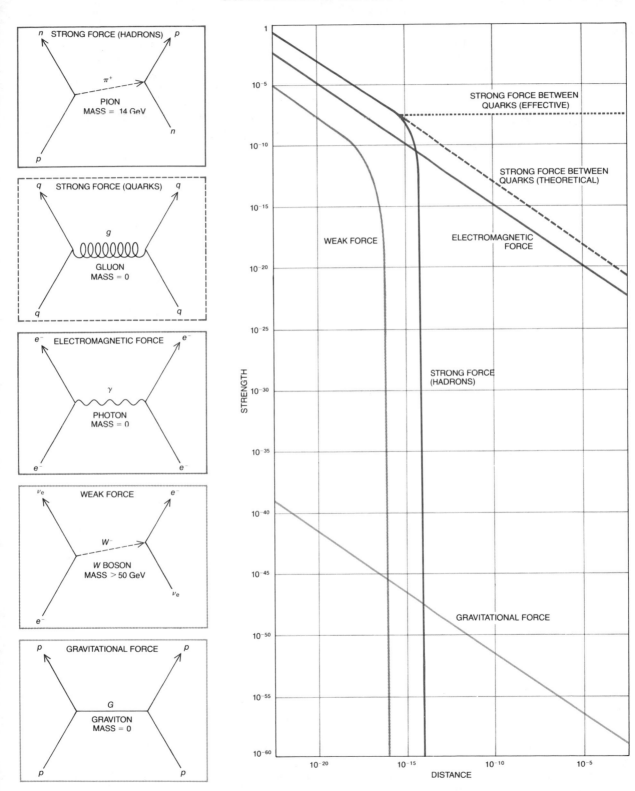

cut from the same blank. The theories in this single favored class are formally designated non-Abelian gauge theories with local symmetry. What is meant by this forbidding label is the main topic of this article. For now, it will suffice to note that the theories relate the properties of the forces to symmetries of nature.

Symmetries and apparent symmetries in the laws of nature have played a part in the construction of physical theories since the time of Galileo and Newton. The most familiar symmetries are spatial or geometric ones. In a snowflake, for example, the presence of a symmetrical pattern can be detected at a glance. The symmetry can be defined as an invariance in the pattern that is observed when some transformation is applied to it. In the case of the snowflake the transformation is a rotation by 60 degrees, or one-sixth of a circle (see Figure 39, top). If the initial position is noted and the snowflake is then turned by 60 degrees (or by any integer multiple of 60 degrees), no change will be perceived. The snowflake is invariant with respect to 60-degree rotations. According to the same principle, a square is invariant with respect to 90-degree rotations and a circle is said to have continuous symmetry because rotation by any angle leaves it unchanged.

Although the concept of symmetry had its origin in geometry, it is general enough to embrace invariance with respect to transformations of other kinds. An example of a nongeometric symmetry is the charge symmetry of electromagnetism. Suppose a number of electrically charged particles have been set out in some definite configuration and all the forces acting between pairs of particles have been measured. If the polarity of all the charges is then reversed, the forces remain unchanged.

Another symmetry of the nongeometric kind concerns isotopic spin, a property of protons and neutrons and of the many related particles called hadrons, which are the only particles responsive to the strong force. The basis of the symmetry lies in the observation that the proton and the neutron are remarkably similar particles. They differ in mass by only about a tenth of a percent, and except for their electric charge they are identical in all other properties. It therefore seems that all protons and neutrons could be interchanged and the strong interactions would hardly be altered. If the electromagnetic forces (which depend on electric charge) could somehow be turned off, the isotopic-spin symmetry would be exact; in reality it is only approximate.

Although the proton and the neutron seem to be distinct particles and it is hard to imagine a state of matter intermediate between them, it turns out that symmetry with respect to isotopic spin is a continuous symmetry, like the symmetry of a sphere rather than like that of a snowflake. I shall give a simplified explanation of why that is so (see Figure 39, bottom). Imagine that inside each particle are a pair of crossed arrows, one representing the proton component of the particle and the other representing the neutron component. If the proton arrow is pointing up (it makes no difference what direction is defined as up), the particle is a proton; if the neutron arrow is up, the particle is a neutron. Intermediate positions correspond to quantum-mechanical superpositions of the two states, and the particle then looks sometimes like a proton and sometimes like a neutron. The symmetry transformation associated with isotopic spin rotates the internal indicators of all protons and neutrons everywhere in the universe by the same amount and at the same time. If the rotation is by exactly 90 degrees, every proton becomes a neutron and every neutron becomes a proton. Symmetry with respect to isotopic spin, to the extent it is exact, states that no effects of this transformation can be detected.

All the symmetries I have discussed so far can be characterized as global symmetries; in this context the word global means "happening everywhere at once." In the description of isotopic-spin symmetry this constraint was made explicit: the internal rotation that transforms protons into neutrons and neutrons into protons is to be carried out everywhere in the universe at the same time. In addition to global symmetries, which are almost always present in a physical theory, it is possible to have a "local" symmetry, in which the convention can be decided independently at every point in space and at every moment in time. Although "local" may suggest something of more modest scope than a global symmetry, in fact the requirement of local symmetry places a far more stringent constraint on the construction of a theory. A global symmetry states that some law of physics remains invariant when the same transformation is applied everywhere at once. For a local symmetry to be observed the law of physics must retain its validity even when a different transformation takes place at each point in space and time.

Gauge theories can be constructed with either a global or a local symmetry (or both), but it is the theories with local symmetry that hold the greatest

SYMMETRY TRANSFORMATION
ROTATION BY 60 DEGREES

GEOMETRIC
SYMMETRY

INTERCHANGE OF ELECTRIC CHARGES

CHARGE
SYMMETRY

ROTATION BY 90 DEGREES IN ABSTRACT INTERNAL SPACE

PROTON

NEUTRON

ISOTOPIC-SPIN
SYMMETRY

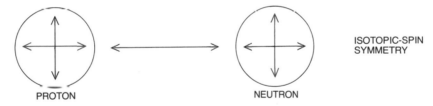

Figure 39 SYMMETRIES OF NATURE. A snowflake pattern is unchanged when it is rotated 60 degrees. Charge symmetry is the invariance of the forces acting among a set of charged particles when the polarities of all the charges are reversed. Isotopic-spin symmetry is based on the observation that little would be changed in the strong interactions of matter if the identities of all protons and neutrons were interchanged. It is symmetries of this kind, where the transformation is an internal rotation or phase shift, that are referred to as gauge symmetries.

interest today. In order to make a theory invariant with respect to a local transformation something new must be added: a force. Before showing how this comes about, however, it will be necessary to discuss in somewhat greater detail how forces are described in modern theories of elementary-particle interactions.

The basic ingredients of particle theory today include not only particles and forces but also fields. A field is simply a quantity defined at every point throughout some region of space and time. For example, the quantity might be temperature and the region might be the surface of a frying pan. The field then consists of temperature values for every point on the surface.

Temperature is called a scalar quantity, because it can be represented by position along a line, or scale. The corresponding temperature field is a scalar field in which each point has associated with it a single number, or magnitude. There are other kinds of field as well, the most important for present purposes being the vector field, where at each point a vector, or arrow, is drawn. A vector has both a magnitude, which is represented by the length of the arrow, and a direction, which in three-dimensional space can be specified by two angles; hence three numbers are needed in order to specify the value of the vector (see Figure 40). An example of a vector field is the velocity field of a fluid; at each point throughout the volume of the fluid an arrow can be drawn to show the speed and direction of flow.

In the physics of electrically charged objects a field is a convenient device for expressing how the force of electromagnetism is conveyed from one place to another. All charged particles are supposed to emanate an electromagnetic field; each particle then interacts with the sum of all the fields rather than directly with the other particles.

In quantum mechanics the particles themselves can be represented as fields. An electron, for example, can be considered a packet of waves with some finite extension in space. Conversely, it is often convenient to represent a quantum-mechanical field as if it were a particle. The interaction of two particles through their interpenetrating fields can then be summed up by saying the two particles exchange a third particle, which is called the quantum of the field. For example, when two electrons, each surrounded by an electromagnetic field, approach each other and bounce apart, they are said to exchange a photon, the quantum of the electromagnetic field.

The exchanged quantum has only an ephemeral existence. Once it has been emitted it must be reabsorbed, either by the same particle or by another one, within a finite period. It cannot keep going indefinitely, and it cannot be detected in an experiment. Entities of this kind are called virtual particles. The larger their energy, the briefer their existence. In effect a virtual particle borrows or embezzles a quantity of energy, but it must repay the debt before the shortage can be noticed.

The range of an interaction is related to the mass of the exchanged quantum. If the field quantum has a large mass, more energy must be borrowed in order to support its existence, and the debt must be repaid sooner lest the discrepancy be discovered. The distance the particle can travel before it must be reabsorbed is thereby reduced and so the corresponding force has a short range. In the special case where the exchanged quantum is massless the range is infinite.

The number of components in a field corresponds to the number of quantum-mechanical states of the field quantum. The number of possible states is in turn related to the intrinsic spin angular momentum of the particle. The spin angular momentum can take on only discrete values; when the magnitude of the spin is measured in fundamental units, it is always an integer or a half integer. Moreover, it is not only the magnitude of the spin that is quantized but also its direction or orientation. (To be more precise, the spin can be defined by a vector parallel to the spin axis, and the projections, or components, of this vector along any direction in space must have values that are integers or half integers.) The number of possible orientations, or spin states, is equal to twice the magnitude of the spin, plus one. Thus a particle with a spin of one-half, such as the electron, has two spin states: the spin can point parallel to the particle's direction of motion or antiparallel to it. A spin-one particle has three orientations, namely parallel, antiparallel and transverse. A spin-zero particle has no spin axis; since all orientations are equivalent, it is said to have just one spin state (see Figure 40, right).

A scalar field, which has just one component (a magnitude), must be represented by a field quantum that also has one component, or in other words by a spin-zero particle. Such particles are therefore called scalar particles. Similarly, a three-component vector field requires a spin-one field quantum with three spin states: a vector particle. The electromagnetic field is a vector field, and the photon, in conformity with these specifications, has a spin of one unit. The gravitational field is a more complicated structure called a tensor and has 10 components; not all of them are independent, however, and the quantum of the field, the graviton, has a spin of two units, which ordinarily corresponds to five spin states.

In the cases of electromagnetism and gravitation one further complication must be taken into account. Since the photon and the graviton are massless, they must always move with the speed of light. Because of their velocity they have a property not shared by particles with a finite mass: the transverse spin states do not exist. Although in some formal sense the photon has three spin states and the graviton has five, in practice only two of the spin states can be detected.

The first gauge theory with local symmetry was the theory of electric and magnetic fields introduced in 1868 by James Clerk Maxwell. The foun-

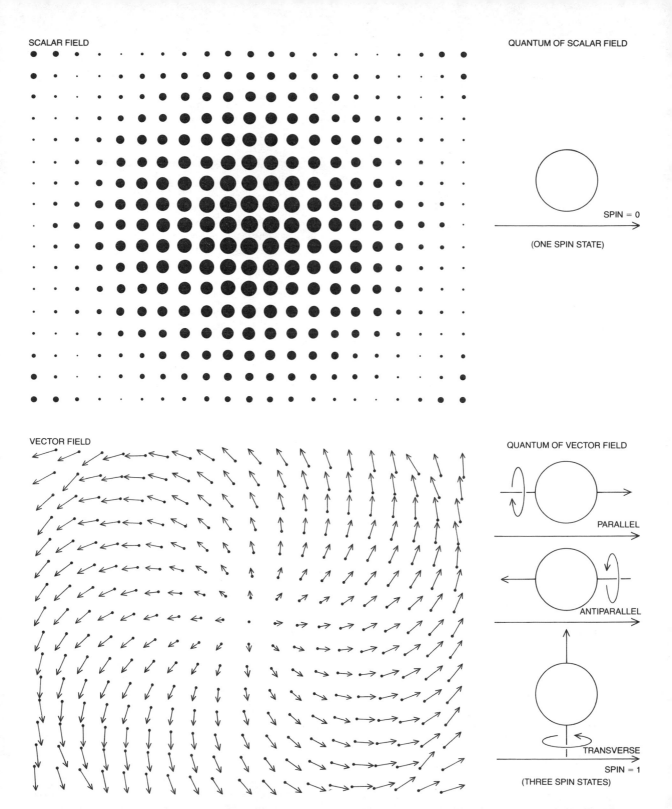

Figure 40 CONCEPT OF A FIELD, a quantity defined at each point throughout some region of space and time, is important in the construction of gauge theories. A scalar field has only a magnitude at each point, given by the area of the dots. A vector field has both a magnitude and a direction, illustrated by drawing an arrow at each point.

dation of Maxwell's theory is the proposition that an electric charge is surrounded by an electric field stretching to infinity, and that the movement of an electric charge gives rise to a magnetic field also of infinite extent. Both fields are vector quantities, being defined at each point in space by a magnitude and a direction.

In Maxwell's theory the value of the electric field at any point is determined ultimately by the distribution of charges around the point. It is often convenient, however, to define a potential, or voltage, that is also determined by the charge distribution: the greater the density of charges in a region, the higher its potential. The electric field between two points is then given by the voltage difference between them.

The character of the symmetry that makes Maxwell's theory a gauge theory can be illustrated by considering an imaginary experiment. Suppose a system of electric charges is set up in a laboratory and the electromagnetic field generated by the charges is measured and its properties are recorded. If the charges are stationary, there can be no magnetic field (since the magnetic field arises from movement of an electric charge); hence the field is purely an electric one. In this experimental situation a global symmetry is readily perceived. The symmetry transformation consists in raising the entire laboratory to a high voltage, or in other words to a high electric potential. If the measurements are then repeated, no change in the electric field will be observed. The reason is that the field, as Maxwell defined it, is determined only by differences in electric potential, not by the absolute value of the potential. It is for the same reason that a squirrel can walk without injury on an uninsulated power line.

This property of Maxwell's theory amounts to a symmetry: the electric field is invariant with respect to the addition or subtraction of an arbitrary overall potential. As noted above, however, the symmetry is a global one, because the result of the experiment remains constant only if the potential is changed everywhere at once. If the potential were raised in one region and not in another, any experiment that crossed the boundary would be affected by the potential difference, just as a squirrel is affected if it touches both a power line and a grounded conductor.

A complete theory of electromagnetic fields must embrace not only static arrays of charges but also moving charges. In order to do that the global symmetry of the theory must be converted into a local symmetry. If the electric field were the only one acting between charged particles, it would not have a local symmetry. Actually when the charges are in motion (but only then), the electric field is not the only one present: the movement itself gives rise to a second field, namely the magnetic field. It is the effects of the magnetic field that restore the local symmetry.

Just as the electric field depends ultimately on the distribution of charges but can conveniently be derived from an electric potential, so the magnetic field is generated by the motion of the charges but is more easily described as resulting from a magnetic potential. It is in this system of potential fields that local transformations can be carried out leaving all the original electric and magnetic fields unaltered. The system of dual, interconnected fields has an exact local symmetry even though the electric field alone does not. Any local change in the electric potential can be combined with a compensating change in the magnetic potential in such a way that the electric and magnetic fields are invariant.

Maxwell's theory of electromagnetism is a classical or non-quantum-mechanical one, but a related symmetry can be demonstrated in the quantum theory of electromagnetic interactions. It is necessary in that theory to describe the electron as a wave or a field, a convention that in quantum mechanics can be adopted for any material particle. It turns out that in the quantum theory of electrons a change in the electric potential entails a change in the phase of the electron wave.

The electron has a spin of one-half unit and so has two spin states (parallel and antiparallel). It follows that the associated field must have two components. Each of the components must be represented by a complex number, that is, a number that has both a real, or ordinary, part and an imaginary part, which includes as a factor the square root of -1. The electron field is a moving packet of waves, which are oscillations in the amplitudes of the real and the imaginary components of the field. It is important to emphasize that this field is not the electric field of the electron but instead is a matter field. It would exist even if the electron had no electric charge. What the field defines is the probability of finding an electron in a specified spin state at a given point and at a given moment. The probability is given by the sum of the squares of the real and the imaginary parts of the field.

In the absence of electromagnetic fields the frequency of the oscillations in the electron field is proportional to the energy of the electron, and the wavelength of the oscillations is proportional to the momentum. In order to define the oscillations completely one additional quantity must be known: the phase. The phase measures the displacement of the wave from some arbitrary reference point and is usually expressed as an angle. If at some point the real part of the oscillation, say, has its maximum positive amplitude, the phase at that point might be assigned the value zero degrees. Where the real part next falls to zero the phase is 90 degrees and where it reaches its negative maximum the phase is 180 degrees. In general the imaginary part of the amplitude is 90 degrees out of phase with the real part, so that whenever one part has a maximal value the other part is zero.

It is apparent that the only way to determine the phase of an electron field is to disentangle the contributions of the real and the imaginary parts of the amplitude. That turns out to be impossible, even in principle. The sum of the squares of the real and the imaginary parts can be known, but there is no way of telling at any given point or at any moment how much of the total derives from the real part and how much from the imaginary part. Indeed, an exact symmetry of the theory implies that the two contributions are indistinguishable. Differences in the phase of the field at two points or at two moments can be measured, but not the absolute phase.

The finding that the phase of an electron wave is inaccessible to measurement has a corollary: the phase cannot have an influence on the outcome of any possible experiment. If it did, that experiment could be used to determine the phase. Hence the electron field exhibits a symmetry with respect to arbitrary changes of phase. Any phase angle can be added to or subtracted from the electron field and the results of all experiments will remain invariant.

This principle can be made clearer by considering an example: the two-slit diffraction experiment with electrons, which is the best-known demonstration of the wavelike nature of matter (see Figures 41 and 42). In the experiment a beam of electrons passes through two narrow slits in a screen and the number of electrons reaching a second screen is counted. The distribution of electrons across the surface of the second screen forms a diffraction pattern of alternating peaks and valleys.

The quantum-mechanical interpretation of this experiment is that the electron wave splits into two

segments on striking the first screen and the two diffracted waves then interfere with each other. Where the waves are in phase the interference is constructive and many electrons are counted at the second screen; where the waves are out of phase destructive interference reduces the count. Clearly it is only the difference in phase that determines the pattern formed. If the phases of both waves were shifted by the same amount, the phase difference at each point would be unaffected and the same pattern of constructive and destructive interference would be observed.

It is symmetries of this kind, where the phase of a quantum field can be adjusted at will, that are called gauge symmetries. Although the absolute value of the phase is irrelevant to the outcome of experiments, in constructing a theory of electrons it is still necessary to specify the phase. The choice of a particular value is called a gauge convention.

Gauge symmetry is not a very descriptive term for such an invariance, but the term has a long history and cannot now be dislodged. It was introduced in about 1920 by Hermann Weyl, who was then attempting to formulate a theory that would combine electromagnetism and the general theory of relativity. Weyl was led to propose a theory that remained invariant with respect to arbitrary dilations or contractions of space. In the theory a separate standard of length and time had to be adopted at every point in space-time. He compared the choice of a scale convention to a choice of gauge blocks, the polished steel blocks employed by machinists as a standard of length. The theory was nearly correct, the necessary emendation being to replace "length scales" by "phase angles." Writing in German, Weyl had referred to "Eich Invarianz," which was initially translated as "calibration invariance," but the alternative translation "gauge" has since become standard.

The symmetry of the electron matter field described above is a global symmetry: the phase of the field must be shifted in the same way everywhere at once. It can easily be demonstrated that a theory of electron fields alone, with no other forms of matter or radiation, is not invariant with respect to a corresponding local gauge transformation. Consider again the two-slit diffraction experiment with electrons. An initial experiment is carried out as before and the electron-diffraction pattern is recorded. Then the experiment is repeated, but one

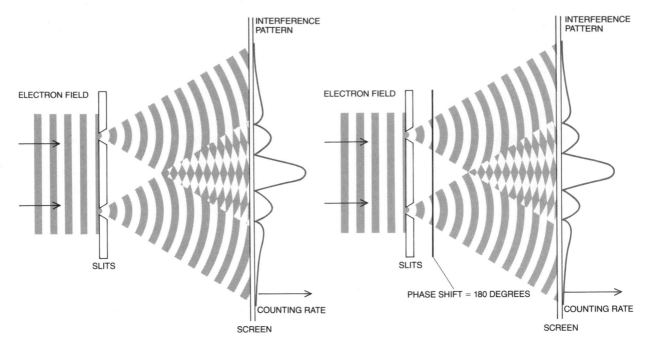

Figure 41 GAUGE SYMMETRY OF ELECTROMAGNET-ISM. When the waves of an electron field pass through a pair of slits, the peaks in the interference pattern are found wherever the waves are in phase; nodes are found where the waves are out of phase. A shift in phase greatly alters the configuration of the field, but it leaves the observable interference pattern unchanged.

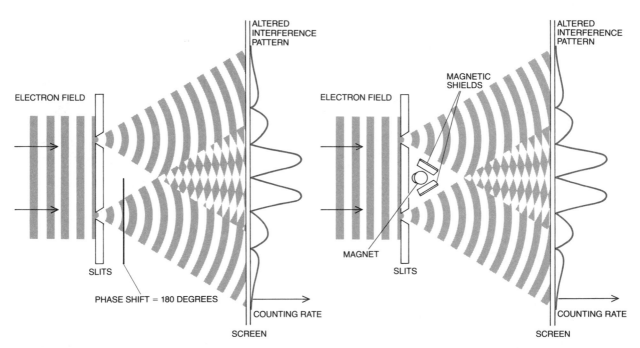

Figure 42 LOCAL GAUGE SYMMETRY of the electron field is restored when magnetic fields are taken into account. Shifting the phase of one diffracted electron beam but not the other clearly alters the observed interference pattern (*left*). The same effect can be obtained by introducing a magnetic field perpendicular to the electron beam and between the slits (*right*). Remarkably, the magnetic field induces the phase shift even when shields are arranged so that the field cannot penetrate the region where the electron waves propagate and interfere.

slit is fitted with the electron-optical equivalent of a half-wave plate, a device that shifts the phase of a wave by 180 degrees. When the waves emanating from the two slits now interfere, the phase difference between them will be altered by 180 degrees. As a result wherever the interference was constructive in the first experiment it will now be destructive, and vice versa. The observed diffraction pattern will not be unchanged; on the contrary, the positions of all the peaks and depressions will be interchanged.

Suppose one wanted to make the theory consistent with a local gauge symmetry. Perhaps it could be fixed in some way; in particular, perhaps another field could be added that would compensate for the changes in electron phase. The new field would of course have to do more than mend the defects in this one experiment. It would have to preserve the invariance of all observable quantities when the phase of the electron field was altered in any way from place to place and from moment to moment. Mathematically the phase shift must be allowed to vary as an arbitrary function of position and time.

Although it may seem improbable, a field can be constructed that meets these specifications. It turns out that the required field is a vector one, corresponding to a field quantum with a spin of one unit. Moreover, the field must have infinite range, since there is no limit to the distance over which the phases of the electron fields might have to be reconciled. The need for infinite range implies that the field quantum must be massless. These are the properties of a field that is already familiar: the electromagnetic field, whose quantum is the photon.

How does the electromagnetic field ensure the gauge invariance of the electron field? It should be remembered that the effect of the electromagnetic field is to transmit forces between charged particles. These forces can alter the state of motion of the particles; what is most important in this context, they can alter the phase. When an electron absorbs or emits a photon, the phase of the electron field is shifted. It was shown above that the electromagnetic field itself exhibits an exact local symmetry; by describing the two fields together the local symmetry can be extended to both of them.

The connection between the two fields lies in the interaction of the electron's charge with the electromagnetic field. Because of this interaction the propagation of an electron matter wave in an electric field can be described properly only if the electric potential is specified. Similarly, to describe an elec-

tron in a magnetic field the magnetic vector potential must be specified. Once these two potentials are assigned definite values the phase of the electron wave is fixed everywhere. The local symmetry of electromagnetism, however, allows the electric potential to be given any arbitrary value, which can be chosen independently at every point and at every moment. For this reason the phase of the electron matter field can also take on any value at any point, but the phase will always be consistent with the convention adopted for the electric and the magnetic potentials.

What this means in the two-slit diffraction experiments is that the effects of an arbitrary shift in the phase of the electron wave can be mimicked by applying an electromagnetic field. For example, the change in the observed interference pattern caused by interposing a half-wave plate in front of one slit could be caused instead by placing the slits between the poles of a magnet (see Figure 42, right). From the resulting pattern it would be impossible to tell which procedure had been followed. Since the gauge conventions for the electric and the magnetic potentials can be chosen locally, so can the phase of the electron field.

The theory that results from combining electron matter fields with electromagnetic fields is called quantum electrodynamics. Formulating the theory and proving its consistency was a labor of some 20 years, begun in the 1920's by P. A. M. Dirac and essentially completed in about 1948 by Richard P. Feynman, Julian Schwinger, Sin-itiro Tomonaga and others.

The symmetry properties of quantum electrodynamics are unquestionably appealing, but the theory can be invested with physical significance only if it agrees with the results of experiments. Indeed, before sensible experimental predictions can even be made the theory must pass certain tests of internal consistency. For example, quantum-mechanical theories predict the probabilities of events: the probabilities must not be negative, and all the probabilities taken together must add up to 1. In addition energies must be assigned positive values but should not be infinite.

It was not immediately apparent that quantum electrodynamics could qualify as a physically acceptable theory. One problem arose repeatedly in an attempt to calculate the result of even the simplest electromagnetic interactions, such as the interaction between two electrons. The likeliest sequence of events in such an encounter is that one electron emits a single virtual photon and the other electron absorbs it. Many more complicated exchanges, however, are also possible; indeed, their number is infinite. For example, the electrons could interact by exchanging two photons, or three, and so on. The total probability of the interaction is determined by the sum of the contributions of all the possible events.

Feynman introduced a systematic procedure for tabulating these contributions by drawing diagrams of the events in one spatial dimension and one time dimension. A notably troublesome class of diagrams are those that include "loops," such as the loop in space-time that is formed when a virtual photon is emitted and later reabsorbed by the same electron. As was shown above, the maximum energy of a virtual particle is limited only by the time needed for it to reach its destination. When a virtual photon is emitted and reabsorbed by the same particle, the distance covered and the time required can be reduced to zero, and so the maximum energy can be infinite. For this reason some diagrams with loops make an infinite contribution to the strength of the interaction.

The infinities encountered in quantum electrodynamics led initially to predictions that have no reasonable interpretation as physical quantities. Every interaction of electrons and photons was assigned an infinite probability. The infinities spoiled even the description of an isolated electron: because the electron can emit and reabsorb virtual particles it has infinite mass and infinite charge.

The cure for this plague of infinities is the procedure called renormalization. Roughly speaking, it works by finding one negative infinity for each possitive infinity, so that in the sum of all the possible contributions the infinities cancel. The achievement of Schwinger and of the other physicists who worked on the problem was to show that a finite residue could be obtained by this method. The finite residue is the theory's prediction. It is uniquely determined by the requirement that all interaction probabilities come out finite and positive.

The rationale of this procedure can be explained as follows. When a measurement is made on an electron, what is actually measured is not the mass or the charge of the pointlike particle with which the theory begins but the properties of the electron together with its enveloping cloud of virtual particles. Only the net mass and charge, the measurable quantities, are required to be finite at all stages of the calculation. The properties of the pointlike ob-

ject, which are called the "bare" mass and the "bare" charge, are not well defined.

Initially it appeared that the bare mass would have to be assigned a value of negative infinity, an absurdity that made many physicists suspicious of the renormalized theory. A more careful analysis, however, has shown that if the bare mass is to have any definite value, it tends to zero. In any case all quantities with implausible values are unobservable, even in principle. Another objection to the theory is more profound: mathematically quantum electrodynamics is not perfect. Because of the methods that must be used for making predictions in the theory the predictions are limited to a finite accuracy of some hundreds of decimal places.

Clearly the logic and the internal consistency of the renormalization method leave something to be desired. Perhaps the best defense of the theory is simply that it works very well. It has yielded results that are in agreement with experiments to an accuracy of about one part in a billion, which makes quantum electrodynamics the most accurate physical theory ever devised. It is the model for the theories of the other fundamental forces and the standard by which such theories are judged.

At the time quantum electrodynamics was completed another theory based on a local gauge symmetry had already been known for some 30 years. It is Einstein's general theory of relativity. The symmetry in question pertains not to a field distributed through space and time but to the structure of space-time itself.

Every point in space-time can be labeled by four numbers, which give its position in the three spatial dimensions and its sequence in the one time dimension. These numbers are the coordinates of the event, and the procedure for assigning such numbers to each point in space-time is a coordinate system. On the earth, for example, the three spatial coordinates are commonly given as longitude, latitude and altitude; the time coordinate can be given in hours past noon. The origin in this coordinate system, the point where all four coordinates have values of zero, lies at noon at sea level where the prime meridian crosses the Equator.

The choice of such a coordinate system is clearly a matter of convention. Ships at sea could navigate just as successfully if the origin of the coordinate system were shifted to Utrecht in the Netherlands. Every point on the earth and every event in its history would have to be assigned new coordinates, but calculations made with those coordinates would invariably give the same results as calculations made in the old system. In particular any calculation of the distance between two points would give the same answer.

The freedom to move the origin of a coordinate system constitutes a symmetry of nature. Actually there are three related symmetries: all the laws of nature remain invariant when the coordinate system is transformed by translation, by rotation or by mirror reflection. It is vital to note, however, that the symmetries are only global ones. Each symmetry transformation can be defined as a formula for finding the new coordinates of a point from the old coordinates. Those formulas must be applied simultaneously in the same way to all the points.

The general theory of relativity stems from the fundamental observation that the structure of space-time is not necessarily consistent with a coordinate system made up entirely of straight lines meeting at right angles; instead a curvilinear coordinate system may be needed. The lines of longitude and latitude employed on the earth constitute such a system, since they follow the curvature of the earth.

In such a system a local coordinate transformation can readily be imagined. Suppose height is defined as vertical distance from the ground rather than from mean sea level. The digging of a pit would then alter the coordinate system, but only at those points directly over the pit. The digging itself represents the local coordinate transformation. It would appear that the laws of physics (or the rules of navigation) do not remain invariant after such a transformation, and in a universe without gravitational forces that would be the case. An airplane set to fly at a constant height would dip suddenly when it flew over the excavation, and the accelerations needed to follow the new profile of the terrain could readily be detected.

As in electrodynamics, local symmetry can be restored only by adding a new field to the theory; in general relativity the field is of course that of gravitation. The presence of this field offers an alternative explanation of the accelerations detected in the airplane: they could result not from a local change in the coordinate grid but from an anomaly in the gravitational field. The source of the anomaly is of no concern: it could be a concentration of mass in the earth or a distant object in space. The point is that any local transformation of the coordinate system could be reproduced by an appropriate set of gravitational fields. The pilot of the airplane could not distinguish one effect from the other.

Both Maxwell's theory of electromagnetism and Einstein's theory of gravitation owe much of their beauty to a local gauge symmetry; their success has long been an inspiration to theoretical physicists. Until recently theoretical accounts of the other two forces in nature have been less satisfactory. A theory of the weak force formulated in the 1930's by Enrico Fermi accounted for some basic features of the weak interaction, but the theory lacked local symmetry. The strong interactions seemed to be a jungle of mysterious fields and resonating particles. It is now clear why it took so long to make sense of these forces: the necessary local gauge theories were not understood.

The first step was taken in 1954 in a theory devised by C. N. Yang and Robert L. Mills, who were then at the Brookhaven National Laboratory. A similar idea was proposed independently at about the same time by R. Shaw of the University of Cambridge. Inspired by the success of the other gauge theories, these theories begin with an established global symmetry and ask what the consequences would be if it were made a local symmetry.

The symmetry at issue in the Yang-Mills theory is isotopic-spin symmetry, the rule stating that the strong interactions of matter remain invariant (or nearly so) when the identities of protons and neutrons are interchanged (see Figure 43). In the global symmetry any rotation of the internal arrows that indicate the isotopic-spin state must be made simultaneously everywhere. Postulating a local symmetry allows the orientation of the arrows to vary independently from place to place and from moment to moment. Rotations of the arrows can depend on any arbitrary function of position and time. The freedom to choose different conventions for the identity of a nuclear particle in different places constitutes a local gauge symmetry.

As in other instances where a global symmetry is converted into a local one, the invariance can be maintained only if something more is added to the theory. Because the Yang-Mills theory is more complicated than earlier gauge theories it turns out that quite a lot more must be added. When isotopic-spin rotations are made arbitrarily from place to place, the laws of physics remain invariant only if six new fields are introduced. They are all vector fields, and they all have infinite range.

The Yang-Mills fields are constructed on the

GLOBAL ISOTOPIC-SPIN ROTATION

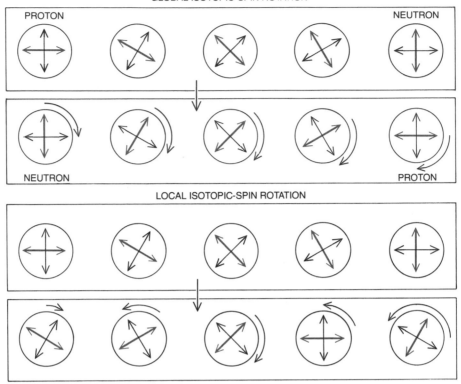

LOCAL ISOTOPIC-SPIN ROTATION

Figure 43 IF ISOTOPIC-SPIN SYMMETRY is valid, the choice of which position of the internal arrow indicates a proton and which a neutron is entirely a matter of convention. Global symmetry (*upper diagram*) requires the same convention to be adopted everywhere, and any rotation of the arrow must be made in the same way at every point. In the Yang-Mills theory isotopic spin is made a local symmetry (*lower diagram*), so that the orientation of the arrow is allowed to vary from place to place.

model of electromagnetism, and indeed two of them can be identified with the ordinary electric and magnetic fields. In other words, they describe the field of the photon. The remaining Yang-Mills fields can also be taken in pairs and interpreted as electric and magnetic fields, but the photons they describe differ in a crucial respect from the known properties of the photon: they are still massless spin-one particles, but they carry an electric charge. One photon is negative and one is positive.

The imposition of an electric charge on a photon has remarkable consequences. The photon is defined as the field quantum that conveys electromagnetic forces from one charged particle to another. If the photon itself has a charge, there can be direct electromagnetic interactions among the pho-

tons. To cite just one example, two photons with opposite charges might bind together to form an "atom" of light. The familiar neutral photon never interacts with itself in this way.

The surprising effects of charged photons become most apparent when a local symmetry transformation is applied more than once to the same particle. In quantum electrodynamics, as was pointed out above, the symmetry operation is a local change in the phase of the electron field, each such phase shift being accompanied by an interaction with the electromagnetic field. It is easy to imagine an electron undergoing two phase shifts in succession, say by emitting a photon and later absorbing one. Intuition suggests that if the sequence of the phase shifts were reversed, so that first a photon was absorbed and later one was emitted, the end result would be the same. This is indeed the case. An unlimited series of phase shifts can be made, and the final result will be simply the algebraic sum of all the shifts no matter what their sequence.

In the Yang-Mills theory, where the symmetry operation is a local rotation of the isotopic-spin arrow, the result of multiple transformations can be quite different. Suppose a hadron is subjected to a gauge transformation, A, followed soon after by a second transformation, B; at the end of this sequence the isotopic-spin arrow is found in the orientation that corresponds to a proton. Now suppose the same transformations were applied to the same hadron but in the reverse sequence: B followed by A. In general the final state will not be the same; the particle may be a neutron instead of a proton. The net effect of the two transformations depends explicitly on the sequence in which they are applied.

Because of this distinction quantum electrodynamics is called an Abelian theory and the Yang-Mills theory is called a non-Abelian one. The terms are borrowed from the mathematical theory of groups and honor Niels Henrik Abel, a Norwegian mathematician who lived in the early years of the 19th century. Abelian groups are made up of transformations that, when they are applied one after another, have the commutative property; non-Abelian groups are not commutative.

Commutation is familiar in arithmetic as a property of addition and multiplication, where for any numbers A and B it can be stated that $A + B = B + A$ and $A \times B = B \times A$. How the principle can be applied to a group of transformations can be illustrated with a familiar example: the group of rotations (see Figure 44). All possible rotations of a two-dimensional object are commutative, and so the group of such rotations is Abelian. For instance, rotations of $+60$ degrees and -90 degrees yield a net rotation of -30 degrees no matter which is applied first. For a three-dimensional object free to rotate about three axes the commutative law does not hold, and the group of three-dimensional rotations is non-Abelian. Consider an airplane heading due north in level flight. A 90-degree yaw to the left followed by a 90-degree roll to the left leaves the airplane heading west with its left wing tip pointing straight down. Reversing the sequence of transformations, so that a 90-degree roll to the left is followed by a 90-degree left yaw, puts the airplane in a nose dive with the wings aligned on the north-south axis.

Like the Yang-Mills theory, the general theory of relativity is non-Abelian: in making two successive coordinate transformations, the order in which they are made usually has an effect on the outcome. In the past 10 years or so several more non-Abelian theories have been devised, and even the electromagnetic interactions have been incorporated into a larger theory that is non-Abelian. For now, at least, it seems all the forces of nature are governed by non-Abelian gauge theories.

ABELIAN TRANSFORMATION

NON-ABELIAN TRANSFORMATION

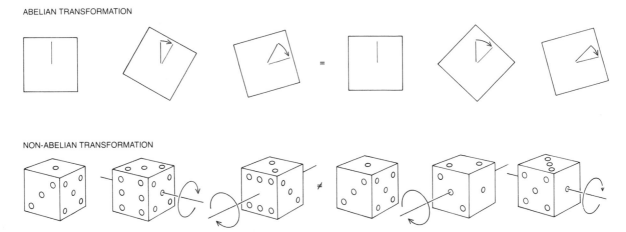

Figure 44 EFFECTS OF REPEATED TRANSFORMATIONS distinguish quantum electrodynamics, an Abelian theory, from the non-Abelian Yang-Mills theory. An Abelian transformation is commutative: if two transformations are applied in succession, the outcome is the same no matter which sequence is chosen (*top*). Non-Abelian transformations are not commutative, so that two transformations will generally yield different results if their sequence is reversed. Rotations in three dimensions exhibit this dependence on sequence (*bottom*).

The Yang-Mills theory has proved to be of monumental importance, but as it was originally formulated it was totally unfit to describe the real world. A first objection to it is that isotopic-spin symmetry becomes exact, with the result that protons and neutrons are indistinguishable; this situation is obviously contrary to fact. Even more troubling is the prediction of electrically charged photons. The photon is necessarily massless because it must have an infinite range. The existence of any electrically charged particle lighter than the electron would alter the world beyond recognition. Of course, no such particle has been observed. In spite of these difficulties the theory has great beauty and philosophical appeal. One strategy adopted in an attempt to fix its defects was to artificially endow the charged field quanta with a mass greater than zero.

Imposing a mass of the quanta of the charged fields does not make the fields disappear, but it does confine them to a finite range. If the mass is large enough, the range can be made as small as is wished. As the long-range effects are removed the existence of the fields can be reconciled with experimental observations. Moreover, the selection of the neutral Yang-Mills field as the only real long-range one automatically distinguishes protons from neutrons. Since this field is simply the electromagnetic field, the proton and the neutron can be distinguished by their differing interactions with it, or in other words by their differing electric charges.

With this modification the local symmetry of the Yang-Mills theory would no longer be exact but approximate, since rotation of the isotopic-spin arrow would now have observable consequences. That is not a fundamental objection: approximate symmetries are quite commonplace in nature. (The bilateral symmetry of the human body is only approximate.) Moreover, at distance scales much smaller than the range of the massive components of the Yang-Mills field, the local symmetry becomes better and better. Thus in a sense the microscopic structure of the theory could remain locally symmetric, but not its predictions of macroscopic, observable events.

The modified Yang-Mills theory was easier to understand, but the theory still had to be given a quantum-mechanical interpretation. The problem of infinities turned out to be severer than it had been in quantum electrodynamics, and the standard recipe for renormalization would not solve it. New techniques had to be devised.

An important idea was introduced in 1963 by Feynman: it is the notion of "ghost" particle, a particle added to a theory in the course of a calculation that vanishes when the calculation is finished. It is known from the outset that the ghost particle is fictitious, but its use can be justified if it never appears in the final state. This can be ensured by making certain the total probability of producing a ghost particle is always zero.

Among theoretical groups that continued work on the Yang-Mills theory the ghost-particle method was taken seriously only at the University of Utrecht, where I was then a student. Martin J. G. Veltman, my thesis adviser, together with John S. Bell of the European Organization for Nuclear Research (CERN) in Geneva, was led to the conclusion that the weak interactions might be described by some form of the Yang-Mills theory. He undertook a systematic analysis of the renormalization problem in the modified Yang-Mills model (with massive charged fields), examining each class of Feynman diagrams in turn. The diagrams having no closed loops were readily shown to make only finite contributions to the total interaction probability. The diagrams with one loop do include infinite terms, but by exploiting the properties of the ghost particles it was possible to make the positive infinities and the negative ones cancel exactly.

As the number of loops increases, the number of diagrams rises steeply; moreover, the calculations required for each diagram become more intricate. To assist in the enormous task of checking all the two-loop diagrams a computer program was written to handle the algebraic manipulation of the probabilities. The output of the program is a list of the coefficients of the infinite quantities remaining after the contributions of all the diagrams have been summed. If the infinities are to be expunged from the theory, the coefficients must without exception be zero. By 1970 the results were known and the possibility of error had been excluded; some infinities remained.

The failure of the modified Yang-Mills theory was to be blamed not on any defect in the Yang-Mills formulation itself but rather on the modifications. The masses of the charged fields had to be put in "by hand" and as a result the invariance with respect to local isotopic-spin rotations was not quite perfect. It was suggested at the time by the Russian investigators L. D. Faddeev, V. N. Popov, E. S. Fradkin and I. V. Tyutin that the pure Yang-Mills theory, with only massless fields, could indeed be renormalized. The trouble with this theory is that

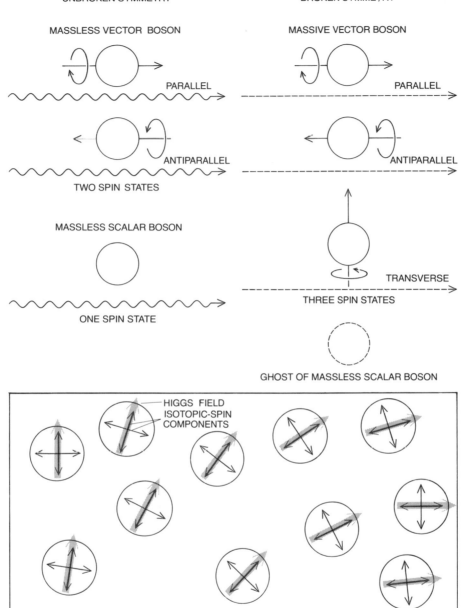

UNBROKEN SYMMETRY

MASSLESS VECTOR BOSON

PARALLEL

ANTIPARALLEL

TWO SPIN STATES

MASSLESS SCALAR BOSON

ONE SPIN STATE

BROKEN SYMMETRY

MASSIVE VECTOR BOSON

PARALLEL

ANTIPARALLEL

TRANSVERSE

THREE SPIN STATES

GHOST OF MASSLESS SCALAR BOSON

HIGGS FIELD
ISOTOPIC-SPIN
COMPONENTS

Figure 45 HIGGS MECHANISM can lend mass to the photonlike vector bosons of the Yang-Mills theory, thereby making the transverse state observable. The Higgs field also provides a frame of reference (gray arrows), which rotates along with the other arrows in a gauge transformation. The relative orientation of the isotopic-spin arrows can be measured with respect to the Higgs arrow.

it not only is unrealistic but also has long-range fields that are difficult to work with.

In the meantime another new ingredient for the formulation of gauge theories had been introduced by F. Englert and Robert H. Brout of the University of Brussels and by Peter Higgs of the University of Edinburgh. They found a way to endow some of the Yang-Mills fields with mass while retaining exact gauge symmetry. The technique is now called the Higgs mechanism (see Figure 45).

The fundamental idea of the Higgs mechanism is to include in the theory an extra field, one having the peculiar property that it does not vanish in the vacuum. One usually thinks of a vacuum as a space with nothing in it, but in physics the vacuum is defined more precisely as the state in which all fields have their lowest possible energy. For most fields the energy is minimized when the value of the field is zero everywhere, or in other words when the field is "turned off." An electron field, for example, has its minimum energy when there are no electrons. The Higgs field is unusual in this respect. Reducing it to zero costs energy; the energy of the field is smallest when the field has some uniform value greater than zero.

The effect of the Higgs field is to provide a frame of reference in which the orientation of the isotopic-spin arrow can be determined. The Higgs field can be represented as an arrow superposed on the other isotopic-spin indicators in the imaginary internal space of a hadron. What distinguishes the arrow of the Higgs field is that it has a fixed length, established by the vacuum value of the field. The orientation of the other isotopic-spin arrows can then be measured with respect to the axis defined by the Higgs field. In this way a proton can be distinguished from a neutron.

It might seem that the introduction of the Higgs field would spoil the gauge symmetry of the theory and thereby lead again to insoluble infinities. In actuality, however, the gauge symmetry is not destroyed but merely concealed. The symmetry specifies that all the laws of physics must remain invariant when the isotopic-spin arrow is rotated in an arbitrary way from place to place. This implies that the absolute orientation of the arrow cannot be determined, since any experiment for measuring the orientation would have to detect some variation in a physical quantity when the arrow was rotated. With the inclusion of the Higgs field the absolute orientation of the arrow still cannot be determined because the arrow representing the Higgs field also rotates during a gauge transformation. All that can be measured is the angle between the arrow of the Higgs field and the other isotopic-spin arrows, or in other words their relative orientations.

The Higgs mechanism is an example of the process called spontaneous symmetry breaking, which was already well established in other areas of physics. The concept was first put forward by Werner Heisenberg in his description of ferromagnetic materials. Heisenberg pointed out that the theory describing a ferromagnet has perfect geometric symmetry in that it gives no special distinction to any one direction in space. When the material becomes magnetized, however, there is one axis—the direction of magnetization—that can be distinguished from all other axes. The theory is symmetrical but the object it describes is not. Similarly, the Yang-Mills theory retains its gauge symmetry with respect to rotations of the isotopic-spin arrow, but the objects described—protons and neutrons—do not express the symmetry.

How does the Higgs mechanism lend mass to the quanta of the Yang-Mills field? The process can be explained as follows. The Higgs field is a scalar quantity, having only a magnitude, and so the quantum of the field must have a spin of zero. The Yang-Mills fields are vectors, like the electromagnetic field, and are presented by spin-one quanta. Ordinarily a particle with a spin of one unit has three spin states (oriented parallel, antiparallel and transverse to its direction of motion), but because the Yang-Mills particles are massless and move with the speed of light they are a special case; their transverse states are missing. If the particles were to acquire a mass, they would lose this special status and all three spin states would have to be observable. In quantum mechanics the accounting of spin states is strict and the extra state must come from somewhere; it comes from the Higgs field. Each Yang-Mills quantum coalesces with one Higgs particle; as a result the Yang-Mills particle gains mass and a spin state, whereas the Higgs particle disappears. A picturesque description of this process has been suggested by Abdus Salam of the International Center for Theoretical Physics in Trieste: the massless Yang-Mills particles "eat" the Higgs particles in order to gain weight, and the swallowed Higgs particles become ghosts.

In 1971, Veltman suggested that I investigate the renormalization of the pure Yang-Mills theory. The rules for constructing the needed Feynman diagrams had already been formulated by Faddeev, Popov, Fradkin and Tyutin, and independently by Bryce S. DeWitt of the University of Texas at Austin and Stanley Mandelstam of the University of California at Berkeley. I could adapt to the task the powerful methods for renormalization studies that had been developed by Veltman.

Formally the results were encouraging, but if the theory was to be a realistic one, some means had to be found to confine the Yang-Mills fields to a finite range. I had just learned at a summer school how Kurt Symanzik of the German Electron Synchrotron and Benjamin W. Lee of the Fermi National Accelerator Laboratory had successfully handled the renormalization of a theoretical model in which a global symmetry is spontaneously broken. It therefore seemed natural to try the Higgs mechanism in the Yang-Mills theory, where the broken symmetry is a local one.

A few simple models gave encouraging results: in these selected instances all infinities canceled no matter how many gauge particles were exchanged and no matter how many loops were included in the Feynman diagrams. The decisive test would come when the theory was checked by the computer program for infinities in all possible diagrams with two loops. The results of that test were available by July, 1971; the output of the program was an uninterrupted string of zeros. Every infinity canceled exactly. Subsequent checks showed that infinities were also absent even in extremely complicated Feynman diagrams. My results were soon confirmed by others, notably by Lee and by Jean Zinn-Justin of the Saclay Nuclear Research Center near Paris.

The Yang-Mills theory had begun as a model of the strong interactions, but by the time it had been renormalized interest in it centered on applications to the weak interactions. In 1967 Steven Weinberg of Harvard University and independently (but later) Salam and John C. Ward of Johns Hopkins University had proposed a model of the weak interactions based on a version of the Yang-Mills theory in which the gauge quanta take on mass through the Higgs mechanism. They speculated that it might be possible to renormalize the theory, but they did not demonstrate it. Their ideas therefore joined many other untested conjectures until some four years later, when my own results showed that it was just that subclass of Yang-Mills theories incorporating the Higgs mechanism that can be renormalized.

The most conspicuous trait of the weak force is its short range: it has a significant influence only to a distance of 10^{-15} centimeter, or roughly a hundredth the radius of a proton. The force is weak largely because its range is so short: particles are unlikely to approach each other closely enough to interact. The short range implies that the virtual particles exchanged in weak interactions must be very massive. Present estimates run to between 80 and 100 times the mass of the proton.

The Weinberg-Salam-Ward model (see Figure 46) actually embraces both the weak force and electromagnetism. It is a local gauge theory. The model applies to the interactions of the particles called leptons, which include the electron (e^-), the muon (μ^-) and two kinds of neutrino (ν_e) and (ν_μ). The conjecture on which the model is ultimately founded is a postulate of local invariance with respect to isotopic spin; in order to preserve that invariance four photonlike fields are introduced, rather than the three of the original Yang-Mills theory. The fourth photon could be identified with some primordial form of electromagnetism. It corresponds to a separate force, which had to be added to the theory without explanation. For this reason the model should not be called a unified field theory. The forces remain distinct; it is their intertwining that makes the model so peculiar.

At the outset all four of the fields in the Weinberg-Salam-Ward model are of infinite range and therefore must be conveyed by massless quanta; one field carries a negative electric charge, one carries a positive charge and the other two fields are neutral. The spontaneous symmetry breaking introduces four Higgs fields, each field represented by a scalar particle. Three of the Higgs fields are swallowed by Yang-Mills particles, so that both of the charged Yang-Mills particles and one of the neutral ones take on a large mass. These particles are collectively named massive intermediate vector bosons, and they are designated W^+, W^- and Z^0. The fourth Yang-Mills particle, which is a neutral one, remains massless: it is the photon of electromagnetism. Of the Higgs particles, the three that lend mass to the Yang-Mills particles become ghosts and are therefore unobservable, but the last Higgs particle is not absorbed, and it should be seen if enough energy is available to produce it.

The most intriguing prediction of the model was the existence of the Z^0, a particle identical with the photon in all respects except mass, which had not been included in any of the earlier, provisional accounts of the weak force. Without the Z^0 any weak interaction would necessarily entail an exchange of electric charge (see Figure 47, top). Events of this kind are called charged-weak-current events. The Z^0 introduced a new kind of weak interaction, a neutral-weak-current event. By exchanging a Z^0, particles would interact without any transfer of charge and could retain their original identities. Neutral weak currents were first observed in 1973 at CERN.

The elaboration of a successful gauge theory of

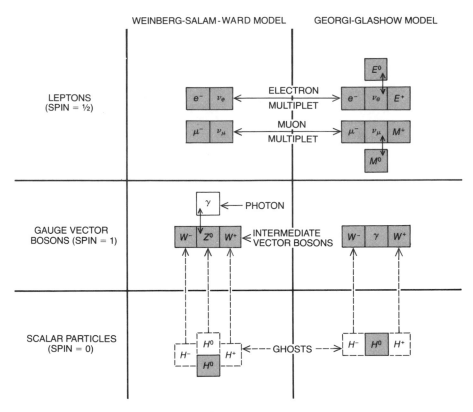

WEINBERG-SALAM-WARD MODEL   GEORGI-GLASHOW MODEL

Figure 46 WEINBERG-SALAM-WARD MODEL applies to the interactions of leptons, which give rise to four massless fields. Three fields are given a mass through the Higgs mechanism and become the intermediate vector bosons. The fourth vector boson is the photon. Three of the Higgs particles are eaten by the vector bosons and become ghosts, but a fourth is left over and should be observable. A theory proposed by Georgi and Glashow suggested a more profound unification, where the photon and the massive vector bosons were in the same family. That theory is now contradicted by experiment.

the strong interactions, which are unique to hadrons, could not be undertaken until a fundamental fact about the hadrons was understood: they are not elementary particles. A model of hadrons as composite objects was proposed in 1963 by Murray Gell-Mann of the California Institute of Technology; a similar idea was introduced independently and at about the same time by Yuval Ne'eman of Tel Aviv University and George Zweig of Cal Tech. In this model hadrons are made up of the smaller particles Gell-Mann named quarks. A hadron can be built out of quarks according to either of two blueprints. Combining three quarks gives rise to a baryon, a class of hadrons that includes the proton and neutron. Binding together one quark and one antiquark makes a meson, a class typified by the pions. Every known hadron can be accounted for as one of these allowed combinations of quarks.

In the original model there were just three kinds of quark, designated "up," "down" and "strange." James D. Bjorken of the Stanford Linear Accelerator Center and Sheldon Lee Glashow of Harvard soon proposed adding a fourth quark bearing a property called charm. In 1971 a beautiful argument by Glashow, John Iliopoulos of Paris and Luciano Maiani of the University of Rome showed that a quark with charm is needed to cure a discrepancy in the gauge theory of weak interactions. Charmed quarks, it was concluded, must exist if both the gauge theory and the quark theory are correct. The discovery in 1974 of the J or psi particle, which consists of a charmed quark and a charmed antiquark, supported the Weinberg-Salam-Ward model and persuaded many physicists that the quark model as a whole should be taken seriously. It now appears that at least two more

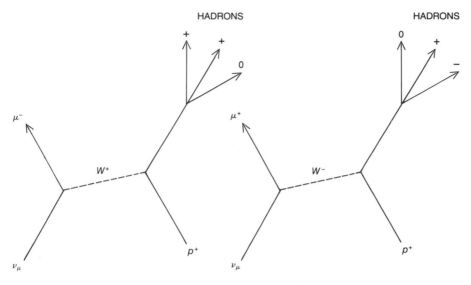

CHARGED WEAK CURRENTS

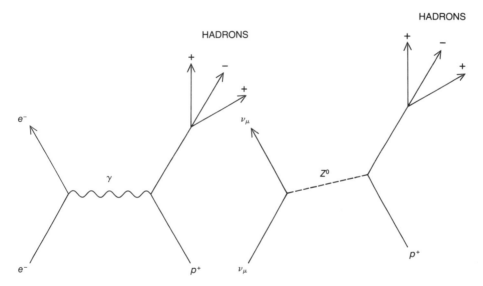

NEUTRAL ELECTROMAGNETIC CURRENT   NEUTRAL WEAK CURRENT

Figure 47 NEUTRAL WEAK CURRENTS provide the decisive test of the Weinberg-Salam-Ward model, which predicts that weak interactions can proceed with or without charge transfer. These neutral weak currents are mediated by the neutral boson Z^0, which is identical with the photon except that it has a very large mass.

"flavors," or kinds, of quark are needed; they have been labeled "top" and "bottom."

The primary task of any theory of the strong interactions is to explain the peculiar rules for building hadrons out of quarks. The structure of a meson is not too difficult to account for: since the meson consists of a quark and an antiquark, it is merely necessary to assume that the quarks carry some property analogous to electric charge. The binding of a quark and an antiquark would then be explained on the principle that opposite charges attract, just as they do in the hydrogen atom. The structure of the baryons, however, is a deeper enigma. To explain how three quarks can form a bound state one must assume that three like charges attract.

The theory that has evolved to explain the strong force prescribes exactly these interactions. The analogue of electric charge is a property called color (although it can have nothing to do with the colors of the visible spectrum). The term color was chosen because the rules for forming hadrons can be expressed succinctly by requiring all allowed combinations of quarks to be "white," or colorless. The quarks are assigned the primary colors red, green and blue; the antiquarks have the complementary "anticolors" cyan, magenta and yellow. Each of the quark flavors comes in all three colors, so that the introduction of the color charge triples the number of distinct quarks (see Figure 48).

From the available quark pigments there are two ways to create white: by mixing all three primary colors or by mixing one primary color with its complementary anticolor. The baryons are made according to the first scheme: the three quarks in a baryon are required to have different colors, so that the three primary hues are necessarily represented. In a meson a color is always accompanied by its complementary anticolor.

The theory devised to account for these baffling interactions is modeled directly on quantum electro-dynamics and is called quantum chromodynamics. It is a non-Abelian gauge theory. The gauge symmetry is an invariance with respect to local transformations of quark color.

It is easy to imagine a global color symmetry (see Figure 49). The quark colors, like the isotopic-spin states of hadrons, might be indicated by the orientation of an arrow in some imaginary internal space. Successive rotations of a third of a turn would change a quark from red to green to blue and back to red again. In a baryon, then, there would be three arrows, with one arrow set to each of the three colors. A global symmetry transformation, by definition, must affect all three arrows in the same way and at the same time. For example, all three arrows might rotate clockwise a third of a turn. As a result of such a transformation all three quarks would change color, but all observable properties of the hadron would remain as before. In particular there would still be one quark of each color, and so the baryon would remain colorless.

Quantum chromodynamics requires that this invariance be retained even when the symmetry transformation is a local one. In the absence of forces or interactions the invariance is obviously lost. Then a local transformation can change the color of one quark but leave the other quarks unaltered, which would give the hadron a net color. As in other gauge theories, the way to restore the in-

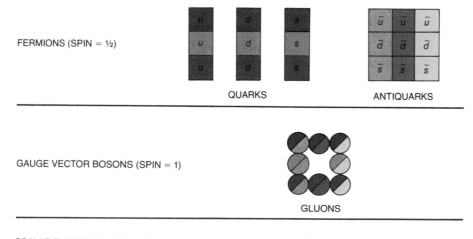

Figure 48 QUARK MODEL. Each quark, labeled u, d or s, has three possible colors, red, green and blue; antiquarks have the corresponding anticolors cyan, magenta and yellow. Interactions of the quarks are described by means of a gauge theory based on invariance with respect to local transformations of color. Sixteen fields are needed to hold this invariance, taken in pairs to make up eight massless vector bosons, called gluons, each bearing a combination of color and anticolor.

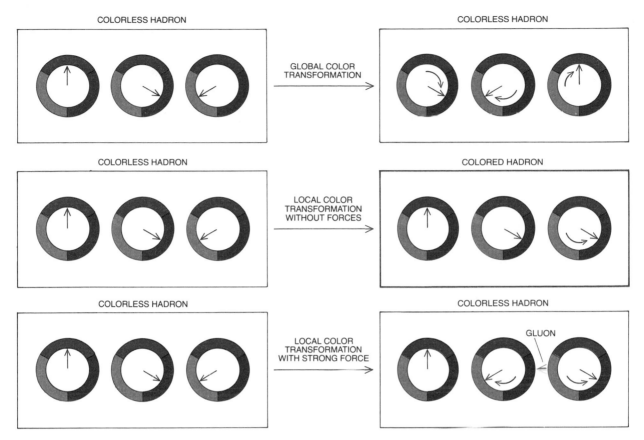

Figure 49 COLOR SYMMETRY requires that every hadron remain colorless, even when the colors of its constituent quarks have been altered. The color of a quark can be indicated by the position of an arrow in an imaginary internal space. In the absence of forces between the quarks, global symmetry cannot be converted into a local symmetry. Changing the position of one color arrow while leaving the other two arrows fixed gives the hadron a net color. In order to preserve the local color symmetry, forces must be introduced.

variance with respect to local symmetry operations is to introduce new fields. In quantum chromodynamics the fields needed are analogous to the electromagnetic field but are much more complicated; they have eight times as many components as the electromagnetic field has. It is these fields that give rise to the strong force.

The quanta of the color fields are called gluons (because they glue the quarks together). There are eight of them, and they are all massless and have a spin angular momentum of one unit. In other words, they are massless vector bosons like the photon. Also like the photon the gluons are electrically neutral, but they are not color-neutral. Each gluon carries one color and one anticolor.

There are nine possible combinations of a color and an anticolor, but one of them is equivalent to white and is excluded, leaving eight distinct gluon fields.

The gluons preserve local color symmetry in the following way. A quark is free to change its color, and it can do so independently of all other quarks, but every color transformation must be accompanied by the emission of a gluon, just as an electron can shift its phase only by emitting a photon. The gluon, propagating at the speed of light, is then absorbed by another quark, which will have its color shifted in exactly the way needed to compensate for the original change. Suppose, for example, a red quark changes its color to green and in the process emits a gluon that bears the colors red and antigreen (see Figure 50). The gluon is then absorbed by a green quark, and in the ensuing reaction

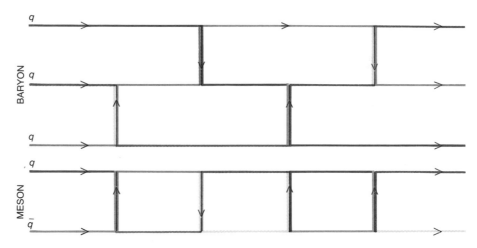

Figure 50 EXCHANGE OF GLUONS maintains a baryon (made up of three quarks) or a meson (made up of a quark and antiquark) colorless. In this process the total color of the particles is conserved.

the green of the quark and the antigreen of the gluon annihilate each other, leaving the second quark with a net color of red. Hence in the final state as in the initial state there is one red quark and one green quark. Because of the continual arbitration of the gluons there can be no net change in the color of a hadron, even though the quark colors vary freely from point to point. All hadrons remain white, and the strong force is nothing more than the system of interactions needed to maintain that condition.

In spite of the complexity of the gluon fields, quantum electrodynamics and quantum chromodynamics are remarkably similar in form. Most notably the photon and the gluon are identical in their spin and in their lack of mass and electric charge. It is curious, then, that the interactions of quarks are very different from those of electrons.

Both electrons and quarks form bound states, namely atoms for the electrons and hadrons for the quarks. Electrons, however, are also observed as independent particles; a small quantity of energy suffices to isolate an electron by ionizing an atom. An isolated quark has never been detected. It seems to be impossible to ionize a hadron, no matter how much energy is supplied. The quarks are evidently bound so tightly that they cannot be pried apart; paradoxically, however, probes of the internal structure of hadrons show the quarks moving freely, as if they were not bound at all.

Gluons too have not been seen directly in experiments. Their very presence in the theory provokes objections like those raised against the pure, massless Yang-Mills theory. If massless particles that so closely resemble the photon existed, they would be easy to detect and they would have been known long ago. Of course, it might be possible to give the gluons a mass through the Higgs mechanism. With eight gluons to be concealed in this way, however, the project becomes rather cumbersome. Moreover, the mass would have to be large or the gluons would have been produced by now in experiments with high-energy accelerators; if the mass is large, however, the range of the quark-binding force becomes too small.

A tentative resolution of this quandary has been discovered not by modifying the color fields but by examining their properties in greater detail. In discussing the renormalization of quantum electrodynamics I pointed out that even an isolated electron is surrounded by a cloud of virtual particles, which it constantly emits and reabsorbs. The virtual particles include not only neutral ones, such as the photon, but also pairs of oppositely charged particles, such as electrons and their antiparticles, the positrons. It is the charged virtual particles in this cloud that under ordinary circumstances conceal the "infinite" negative bare charge of the electron. In the vicinity of the bare charge the electron-positron pairs become slightly polarized: the virtual positrons, under the attractive influence of the bare charge, stay closer to it on the average than the virtual electrons, which are repelled. As a result the bare charge is partially neutralized; what is seen at long range is the difference between the bare charge and the screening charge of the virtual positrons (see Figure 51, left). Only when a probe approaches to within less than about 10^{-10} centimeter do the unscreened effects of the bare charge become significant.

It is reasonable to suppose the same process would operate among color charges, and indeed it does. A red quark is enveloped by pairs of quarks and antiquarks, and the antired charges in this cloud are attracted to the central quark and tend to screen its charge. In quantum chromodynamics, however, there is a competing effect that is not present in quantum electrodynamics. Whereas the photon carries no electric charge and therefore has no direct influence on the screening of electrons, gluons do bear a color charge. (This distinction expresses the fact that quantum electrodynamics is an Abelian theory and quantum chromodynamics is a non-Abelian one.) Virtual gluon pairs also form a cloud around a colored quark, but it turns out that the gluons tend to enhance the color charge rather than attenuate it (Figure 51, right). It is as if the red component of a gluon were attracted to a red quark and therefore added its charge to the total effective charge. If there are no more than 16 flavors of quark (and at present only six are known), the "antiscreening" by gluons is the dominant influence.

This curious behavior of the gluons follows from rather involved calculations, and the interpretation of the results depends on how the calculation was done. When I calculate it, I find that the force responsible is the color analogue of the gluon's magnetic field. It is also significant, however, that virtual gluons can be emitted singly, whereas virtual quarks always appear as a quark and an antiquark. A single gluon, bearing a new color charge, enhances the force acting between two other color charges.

As a result of this "antiscreening" the effective color charge of a quark grows larger at long range than it is close by. A distant quark reacts to the combined fields of the central quark and the reinforcing gluon charges; at close range, once the gluon cloud has been penetrated, only the smaller bare charge is effective. The quarks in a hadron therefore act somewhat as if they were connected by rubber bands: at very close range, where the bands are slack, the quarks move almost independently, but at a greater distance, where the bands are stretched taut, the quarks are tightly bound.

The polarization of virtual gluons leads to a reasonably precise account of the close-range behavior of quarks. Where the binding is weak, the expected motion of the particles can be calculated successfully. The long-range interactions, and most notably the failure of quarks and gluons to appear as free particles, can probably be attributed to the same mechanism of gluon antiscreening. It seems likely

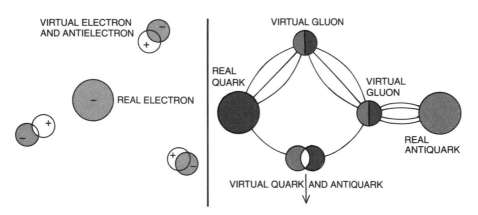

Figure 51 POLARIZATION OF THE VACUUM. Pairs of virtual electrons and antielectrons surround any isolated charge, such as an electron (*left*). The positively charged antielectrons tend to remain nearer the negative electron charge and thereby cancel part of it. The observed electron charge is the difference between the "bare" charge and the screening charge of virtual antielectrons. In quantum chromodynamics pairs of virtual quarks (*right*) diminish the strength of the force between a real quark and a real antiquark, but there is a competing effect—virtual gluons do not screen the quark charge but enhance it.

that as two color charges are pulled apart the force between them grows stronger indefinitely, so that infinite energy would be needed to create a macroscopic separation. This phenomenon of permanent quark confinement may be linked to certain special mathematical properties of the gauge theory. It is encouraging that permanent confinement has indeed been found in some highly simplified models of the theory. In the full-scale theory all methods of calculation fail when the forces become very large, but the principle seems sound. Quarks and gluons may therefore be permanently confined in hadrons.

If the prevailing version of quantum chromodynamics turns out be correct, color symmetry is an exact symmetry and the colors of particles are completely indistinguishable. The theory is a pure gauge theory of the kind first proposed by Yang and Mills. The gauge fields are inherently long-range and formally are much like the photon field. The quantum-mechanical constraints on those fields are so strong, however, that the observed interactions are quite unlike those of electromagnetism and even lead to the imprisonment of an entire class of particles.

Even where the gauge theories are right they are not always useful. The calculations that must be done to predict the result of an experiment are tedious, and except in quantum electrodynamics high accuracy can rarely be attained. It is mainly for practical or technical reasons such as these that the problem of quark confinement has not been solved. The equations that describe a proton in terms of quarks and gluons are about as complicated as the equations that describe a nucleus of medium size in terms of protons and neutrons. Neither set of equations can be solved rigorously.

In spite of these limitations the gauge theories have made an enormous contribution to the understanding of elementary particles and their interactions. What is most significant is not the philosophical appeal of the principle of local symmetry, or even the success of the individual theories. Rather it is the growing conviction that the class of theories now under consideration includes all possible theories for any system of particles whose mutual interactions are not too strong. Experiment shows that if particles remain closer together than about 10^{-14} centimeter, their total interaction, including the effects of all forces whether known or not, is indeed small. (The quarks are a special case: although the interactions between them are not small, those interactions can be attributed to the effects of

virtual particles, and the interactions of the virtual particles are only moderate.) Hence it seems reasonable to attempt a systematic fitting of the existing gauge theories to experimental data.

The mathematics of the gauge theories is rigid, but it does leave some freedom for adjustment. That is, the predicted magnitude of an interaction between particles depends not only on the structure of the theory but also on the values assigned to certain free parameters, which must be considered constants of nature. The theory remains consistent no matter what choice is made for these constants, but the experimental predictions depend strongly on what values are assigned to the constants. Although the constants can be measured by doing experiments, they can never be derived from the theory. Examples of such constants of nature are the charge of the electron and the masses of elementary particles such as the electron and the quarks.

The strength of the gauge theories is that they require comparatively few such free parameters: about 18 constants of nature must be supplied to account for all the known forces. The tangled phenomena of the strongly interacting particles, which seemed incomprehensible 15 years ago, can now be unraveled by means of a theory that includes only a handful of free parameters. Among these all but three are small enough to be safely ignored.

Even if the free parameters have been reduced to a manageable number, they remain an essential part of the theory. No explanation can be offered of why they assume the values they do. The fundamental questions that remain unanswered by the gauge theories center on these apparent constants of nature. Why do the quarks and the other elementary particles have the masses they do? What determines the mass of the Higgs particle? What determines the fundamental unit of electric charge or the strength of the color force? The answers to such questions cannot come from the existing gauge theories but only from a more comprehensive theory.

In the search for a larger theory it is natural to apply once more a recipe that has already proved successful. Hence the obvious program is to search for global symmetries and explore the consequences of making them local symmetries. This principle is not a necessary one, but it is worth trying. Just as Maxwell's theory combined electricity and magnetism and the Weinberg-Salam-Ward model linked electromagnetism and the weak force, so perhaps some larger theory could be found to embrace both

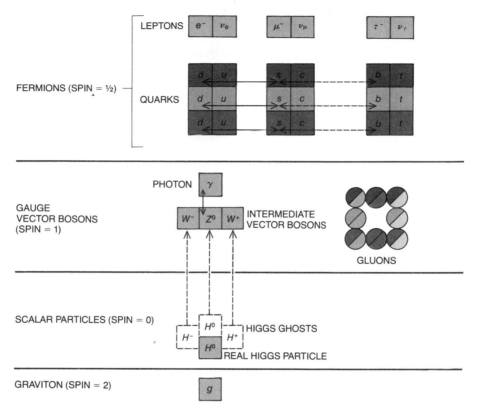

Figure 52 STANDARD MODEL of elementary-particle interactions describes the four forces of nature by means of three non-Abelian gauge theories. The fundamental particles of matter are six leptons and six flavors of quark present in three colors. Electromagnetism and the weak force are mediated by the gauge particles of the Weinberg-Salam-Ward model. The strong force is attributed to the eight massless gluons. Gravitation results from the exchange of a massless spin-2 particle, the graviton. There is one surviving Higgs particle, which is massive and electrically neutral.

the Weinberg-Salam-Ward model and quantum chromodynamics. Such a theory might in principle be constructed on the model of the existing gauge theories. A more sweeping symmetry of nature must be found; making this symmetry a local one would then give rise to the strong force, the weak force and electromagnetism. In the bargain yet more forces, exceedingly weak and so far unobserved, may be introduced.

Work on such theories is proceeding, and it has lately concentrated on symmetries that allow transformations between quarks and leptons, the class of particles that includes the electron. It is my belief the schemes proposed so far are not convincing. The grand symmetry they presuppose must be broken in order to account for the observed disparities among the forces, and that requires several Higgs fields. The resulting theory has as many arbitrary constants of nature as the less comprehensive theories it replaces.

A quite different and more ambitious approach to unification has recently been introduced under the terms "supersymmetry" and "supergravity." It gathers into a single category particles with various quantities of angular momentum; up to now particles with different spins were always assigned to separate categories. The utility of the supersymmetric theories has yet to be demonstrated, but they hold much promise. They offer a highly restrictive description of some hundreds of particles, including the graviton, in terms of only a few adjustable parameters. So far the results do not much resemble the known physical world, but that was also true of the first Yang-Mills theory in 1954.

The form of unification that has been sought longest and most ardently is a reconciliation of the various quantum field theories with the general theory of relativity. The gravitational field seems to lead inevitably to quantized theories that cannot be renormalized. At extremely small scales of distance (10^{-33} centimeter) and time (10^{-44} second) quantum fluctuations of space-time itself become important, and they call into question the very meaning of a space-time continuum. Here lie the present limits not merely of gauge theories but of all known physical theories.

POSTSCRIPT

Even though protons, neutrons and electrons are not truly "elementary particles," attempts to cut them into even smaller pieces are complicated because the velocities at which they are produced and interact approach the speed of light. Relativistic moving particles are more massive than when at rest, energy being converted into mass according to $E = mc^2$. As a consequence, any attempt to subdivide the elementary particles produces "constituents" heavier than the originals.

Not only are the laws of relativity obeyed but also those of quantum mechanics. These two theories, which are mathematically quite different, were combined in quantum field theory. Space and time are flooded with many different fields, of which the familiar electric and magnetic fields were the first and are the best known. There are as many fields as particles and by studying the "quantized" equations of a field the properties of the corresponding particle are found. The electromagnetic field, for instance, defines the properties of photons.

By the early 1970's, we had a complete picture of all possible quantum field theories that could describe elementary particles. Symmetry plays an essential role here: given any symmetry structure there is a corresponding theory for elementary particles, most often a local gauge symmetry. The ensuing theory is a gauge theory, as described in my article.

The surprising discovery of the J/ψ particle in 1974 was a dramatic confirmation of what is now known as the "standard model." The gauge theory description of strong forces is the most difficult because these forces are extremely nonlinear. But by the end of the decade there was little doubt of the basic correctness of quantum chromodynamics. This gauge theory, based on a symmetry called

$SU(3)$, explains how protons and neutrons, like most other particles, are built from more fundamental quarks. It also showed that it is impossible to isolate single quarks because the energy needed to do so would be infinite. Since my article appeared there have been a number of important discoveries. For example, in 1983 the electroweak theory received striking confirmation when experimenters at CERN observed the W^+, W^- and Z^0 with exactly the properties discussed.

My article appeared after several of the points about quantum chromodynamics had become clear, but explanations and descriptions in it need hardly any revision, although some communication difficulties between author and artist, across an ocean, resulted in the figures being somewhat inaccurate. Unlike political and social theories, once a truth has been found in physics, it remains true forever.

Theorists, however, have not been idle since 1980. In gauge theories themselves, most advances have been made at a level of understanding that goes beyond the scope of my article. For instance the details of the mechanism that keep quarks imprisoned in hadronic particles via quantum chromodynamics are qualitatively well understood. A quantitative confirmation requires extremely powerful computer calculations that are still being carried out.

Gauge theories have a rich topological structure. The Higgs field, for example, can be represented by an arrow with a more or less fixed length determined by a balance of different energies. Imagine that this Higgs arrow is arranged in a pattern like that of the skin of a hedgehog rolled up in a defensive posture, the arrows pointing everywhere outwards. "Topology" tells us that if we measure the arrow everywhere, outside and inside the hedgehog, and if the field is continuous as it mostly has to be, then there is a problem: In at least one spot in the hedgehog's intestines the arrow must vanish, and this costs energy. What is obtained is a new model for an exotic particle, the magnetic monopole, a still undiscovered particle carrying a single magnetic charge not accompanied by a nearby countercharge as in usual magnets. These objects are described in Chapter 8, "Superheavy Magnetic Monopoles," by Richard A. Carrigan, Jr. and W. Peter Trower. Another topological feature of gauge theories causes the anomalously fast decay of neutral pions into pairs of photons, while others are at the root of the quark confining mechanism.

The fundamental problems mentioned at the end

of my article are still open and much discussed. Like all theories of nature, gauge theories have a limited applicability. For a particle having so much energy that the gravitational forces come into play, quantum field theory in its present form becomes useless since it would predict that curvature of space-time itself tends to infinity.

One new avenue must be mentioned, the superstring theory. String theories were born in an attempt to understand the strong interaction and the confinement of quarks before quantum chromodynamics was proposed. Quarks were then believed to be held together by strings, but constraints of quantum mechanics and relativity are so restrictive that this approach to understanding quark dynamics was abandoned. String theory is the only candidate theory that could encompass both gauge and gravitation theories, yet it is neither of these. String theory does not describe the forces between quarks, but operates in quite a different world, distances of 10^{-33} centimeter and times of 10^{-44} seconds. In this picture the gravitational force results from exchanges of tiny closed strings.

This subject has gained tremendous popularity since Michael Green [see "Superstrings" SCIENTIFIC AMERICAN, September 1986] and John Schwartz recently showed how to remove some nasty mathematical discrepancies called "anomalies" from string theory in a construct called a "superstring." In some sense string theory is a superlative of gauge theory, where the gauge invariance is the freedom to choose coordinates on a one-dimensional piece of string. This "symmetry" is much more complex than ordinary gauge theory, and is mathematically extremely interesting.

Another fundamental attempt to resolve the shortcomings of gauge theories occurs when elementary particles are compared with black holes, those esoteric space-time abnormalities that remain after a large amount of matter collapses under its own weight. If we understood the quantum properties of a black hole better, elementary particles might look like black pinholes. A unification between elementary particles and black holes would then yield a complete theory of all particles and forces.

UNIFICATION UNOBSERVED

. . .

The Decay of the Proton

The proton is known to have a lifetime at least 10^{20} times the age of the universe, but theory indicates that it may not live forever. If it is not immortal, all ordinary matter will ultimately disintegrate.

. . .

Steven Weinberg

The discovery of radioactivity by Antoine Henri Becquerel in 1896 dispelled the belief that all atoms are permanent and immutable. The energetic particles that had been detected by Becquerel were later understood as being emitted when the nuclei of the atoms of a radioactive substance decay spontaneously into other atomic nuclei. Interesting as this nuclear instability was, it seemed to be a rarity, a property only of certain heavy elements such as uranium and radium. The nuclei of common elements such as hydrogen and oxygen were thought to be absolutely stable.

There are now several theoretical reasons to suspect that all atomic nuclei ultimately decay and hence that all matter is in some small degree radioactive. In a decay of this kind one of the two types of particle in the atomic nucleus, a proton or a neutron, would be spontaneously transformed into energetic particles very different from the particles that make up ordinary atoms. Even the lightest nucleus, that of hydrogen, which consists of a single proton, would be subject to decay.

Much evidence, beginning with the great age of the earth, indicates that matter cannot be highly evanescent. If ordinary matter decays, it does so only very slowly, so slowly that experiments of an extraordinarily large scale will be needed to detect the decay. Becquerel discovered the radioactive disintegration of uranium nuclei in a crystal of uranium salts that weighed perhaps a few grams; in order to observe the feebler radioactivity associated with the decay of the proton, it will be necessary to monitor many tons of material. Nevertheless, experimental searches for the decay of the proton are now under way.

To see what is at stake in these experiments, it is useful first to ask why anything in the world should last forever. The electron, for example, is still thought to be absolutely stable. What physical principles prevent it from decaying into other particles? By understanding the stability of particles such as the electron one can judge whether there are any physical principles that prevent the decay of ordinary atomic nuclei.

Experience in the physics of elementary particles teaches that any decay process one can imagine will occur spontaneously unless it is forbidden by one of the conservation laws of physics. A conservation law states that the total value of some quantity, such as energy or electric charge, can never change. Even if a decay process is not produced directly by one of the fundamental interactions of elementary particles, if it is not forbidden by a conservation law, it will be produced by some more or less complicated sequence of emissions and absorptions of particles. Thus in considering whether any particle is stable one has to ask whether its decay would violate any conservation law.

The law of conservation of energy is easy to apply. It simply requires that the mass of the decaying particle (or the energy equivalent of the mass) be greater than the total mass of the decay products. (It is not enough for the masses to be equal because some mass must be converted into the kinetic energy of the decay products.) Therefore a good way to start in judging the stability of any particle is to list all the less massive particles into which it might conceivably decay.

Consider the electron. As far as is known, there are only a few kinds of particle with a mass less than that of an electron. The most familiar of them is the photon, the quantum of light, whose mass is thought to be exactly zero. There are strong theoretical grounds for thinking there is also a quantum of gravitational radiation, the graviton, again with zero mass. Finally there are various species of particles called neutrinos, which are similar in some respects to the electron; they are emitted in the familiar kind of radioactivity known as beta decay, the kind that was discovered by Becquerel in 1896. Neutrinos have generally been thought to have zero mass, but the determination of their mass is at present an object of intense theoretical and experimental effort. Nevertheless, there is no doubt that at least one of the species of neutrino has a mass less than about a ten-thousandth of the mass of the electron.

Why then does the electron not decay into, say, neutrinos and photons? The answer is that although such a decay would satisfy the law of conservation of energy, it would violate another conservation law, that of electric charge. Benjamin Franklin was the first to recognize that the net quantity of electric charge (positive minus negative) never increases or decreases, although charges of opposite sign can be separated or recombined. Electrons carry a definite negative electric charge, but all the lighter particles into which the electron might decay (the photon, the graviton and the neutrinos) happen to carry zero electric charge. The decay of an electron would entail the destruction of a definite negative quantity of electric charge and is therefore strictly forbidden.

Now consider how these conservation laws might apply to the decay of the two kinds of particle that make up the atomic nucleus. For the moment consider only the lighter of the two particles, the pro-

ton, and return to the neutron later on. The proton carries a positive electric charge, equal in magnitude but opposite in sign to that of the electron, and so it too cannot decay into neutrinos, photons or gravitons. The proton, however, is about 1,820 times as heavy as the electron, and there are several particles of lesser mass that also have positive charge. The proton could decay into these other particles without violating the conservation of either energy or electric charge. For example, the electron has an antiparticle called the positron, with the same mass as the electron but with a positive electric charge equal to that of the proton. (For every kind of particle there is an antiparticle with the same mass but opposite values of other properties, such as electric charge. Incidentally, the positron is stable for the same reason that the electron is.) There is nothing in the laws of energy or charge conservation that would forbid a proton from decaying into a positron and any number of photons and neutrinos.

Another candidate for a decay product of the proton is the antimuon. The muon is a particle similar in many respects to the electron and with the same charge, but it is 210 times as massive. (The muon does decay into an electron and neutrinos.) The antimuon has the same charge as the proton but only about one-ninth the mass. A proton might therefore decay into an antimuon plus light neutral particles such as photons and neutrinos.

Still another possible product of the proton's decay is a meson, a member of the group of unstable particles intermediate in mass between the electron and the proton. The conservation laws of energy and charge would allow the proton to decay into, say, a positively charged meson and a neutrino or into a neutral meson and a positron. Any of these decay processes would lead to the total disruption of the hydrogen atom. In a heavier element they would change the chemical nature of the element and would release energy in amounts much greater than those released in ordinary radioactivity.

Why is matter everywhere not observed to be disintegrating as a result of such decay processes? This problem seems to have been first addressed by Hermann Weyl in 1929. Positrons, muons and mesons were unknown then, and so the conjectural proton-decay schemes outlined above could not have been imagined. Weyl was nonetheless puzzled about the stability of matter; he may have wondered why the protons in an atom do not absorb the orbiting electrons, leading for example to the decay of a hydrogen atom into a shower of photons. Weyl suggested that the stability of matter might be explained if there were two kinds of electric charge, one carried by the electron and the other by the proton. If each kind of charge were conserved separately, the mutual annihilation of a proton and an electron would be forbidden. Weyl's proposal did not attract much attention at the time.

The question was taken up again by E. C. G. Stueckelberg in 1938 and by Eugene P. Wigner (in a footnote) in 1949. They proposed what became the conventional view, namely that in addition to energy and electric charge there is another conserved property of matter, which has since come to be called baryon number. The baryons (from the Greek *barys*, heavy) are a family of particles that includes the proton and many particles heavier than the proton, such as the neutron and the highly unstable particles called hyperons. All baryons are assigned a baryon number of $+1$, and all lighter particles, including the photon, the electron, the positron, the graviton, the neutrino, the muon and the mesons, have a baryon number of zero. For an atom or another composite system of particles the baryon number is the sum of the baryon numbers of the constituent particles. It follows that any collection of particles lighter than the proton has a baryon number of zero. The law of baryon-number conservation is the assertion that the total baryon number cannot change. The decay of a proton into a collection of lighter particles would entail the conversion of a state whose baryon number is $+1$ into a state whose baryon number is zero, and so the decay is forbidden.

An antiparticle has a baryon number opposite to that of the corresponding particle. The antiproton, for example, has a baryon number of -1; it is an antibaryon. A proton and an antiproton can annihilate each other without violating the conservation of baryon number; the proton and the antiproton have a total baryon number of $+1$ plus -1, or zero, and so they can turn into a shower of mesons or photons. Thus baryon-number conservation does not require that each proton be immortal but rather requires that protons not decay spontaneously in ordinary matter, where there are no antiprotons.

So far I have discussed only the decay of the proton, but of course the nucleus of most atoms is made up not only of protons but also of neutrons. What about the possibility of the neutron's decaying? The neutron is a baryon with an electric charge of zero and a mass slightly larger than that of the proton. To be more precise, the mass of the neutron is a little greater than the mass of the proton plus the mass of the electron. This relation suggests one possible mode of decay for the neutron: it could give rise to a proton, an electron and some massless neutral particles. Energy can evidently be conserved in this process. So can electric charge, since the charges of the proton and the electron cancel each other. Baryon number is also conserved, since the neutron and the proton each have a baryon number of $+1$ and the other particles have a baryon number of zero.

A free neutron (one that is not bound in an atomic nucleus) does decay in exactly this matter: it yields a proton, an electron and an antineutrino. The half-life of the free neutron, which is the time required for half of the neutrons in any large sample to decay, is roughly 10 minutes. Neutrons in certain atomic nuclei, such as the nucleus of tritium (the heavy isotope of hydrogen with one proton and two neutrons), can also decay into protons; this is the process of beta decay. In most nuclei, however, neutrons do not decay because too much energy would be required to create a proton amid the repulsive electrostatic forces generated by the other protons in the nucleus. In such nuclei the neutrons are as stable as the protons.

The possibility remains that a neutron bound in a nucleus might decay in some other manner that does not conserve baryon number. For example, it might give rise to a positron and a negatively charged meson or to an electron and a positively charged meson. The discovery of such a neutron decay in an otherwise stable nucleus would be as significant as the discovery of the decay of a proton. Indeed, the experiments that are searching for proton decay are also looking for the decay of bound neutrons. Since the decay of free neutrons is already well known, however, the experimental tests of baryon-number conservation have come to be known as proton-decay experiments.

In recent years it has become widely accepted that the baryons and the mesons are made up of the more fundamental particles called quarks. A baryon consists of three quarks, an antibaryon consists of three antiquarks and a meson consists of a quark and an antiquark. The electron, the muon and neutrinos belong to the family of particles called leptons, which are not made up of quarks and indeed give no sign of an inner structure. On this basis the baryon number of any system of particles is just

PARTICLE	MASS (MeV)	ELECTRIC CHARGE	BARYON NUMBER	PRINCIPAL DECAY MODE
PHOTON (γ)	0	0	0	NONE KNOWN
NEUTRINO (ν)	0?	0	0	NONE KNOWN
ANTINEUTRINO ($\bar{\nu}$)	0?	0	0	NONE KNOWN
ELECTRON (e^-)	.511	−1	0	NONE KNOWN
POSITRON (e^+)	.511	+1	0	NONE KNOWN
MUON (μ^-)	105.7	−1	0	$\mu^- \rightarrow e^- + \nu + \bar{\nu}$ MASS: $105.7 \rightarrow .511 + 0 + 0$ CHARGE: $-1 \rightarrow -1 + 0 + 0$ BARYON NUMBER: $0 \rightarrow 0 + 0 + 0$
ANTIMUON (μ^+)	105.7	+1	0	$\mu^+ \rightarrow e^+ + \nu + \bar{\nu}$ MASS: $105.7 \rightarrow .511 + 0 + 0$ CHARGE: $+1 \rightarrow +1 + 0 + 0$ BARYON NUMBER: $0 \rightarrow 0 + 0 + 0$
PI MESONS (π^+)	139.6	+1	0	$\pi^+ \rightarrow \mu^+ + \nu$ MASS: $139.6 \rightarrow 105.7 + 0$ CHARGE: $+1 \rightarrow +1 + 0$ BARYON NUMBER: $0 \rightarrow 0 + 0$
(π^0)	135	0	0	$\pi^0 \rightarrow \gamma + \gamma$ MASS: $135 \rightarrow 0 + 0$ CHARGE: $0 \rightarrow 0 + 0$ BARYON NUMBER: $0 \rightarrow 0 + 0$
(π^-)	139.6	−1	0	$\pi^- \rightarrow \mu^- + \bar{\nu}$ MASS: $139.6 \rightarrow 105.7 + 0$ CHARGE: $-1 \rightarrow -1 + 0$ BARYON NUMBER: $0 \rightarrow 0 + 0$
K MESONS (K^+)	493.7	+1	0	$K^+ \rightarrow \mu^+ + \nu$ MASS: $493.7 \rightarrow 105.7 + 0$ CHARGE: $+1 \rightarrow + 1 + 0$ BARYON NUMBER: $0 \rightarrow 0 + 0$

Figure 53 PRINCIPAL DECAY MODES of particles are listed in order of increasing mass. The mass is given in terms of equivalent energy in units of a million electron volts (MeV). The charge is given in units of the charge of the proton. Under decay mode is an accounting of the three conserved quantities. The decay of a proton into a collection of lighter particles would entail a transformation of a state whose baryon number is +1 into a state whose baryon number is zero, and so the decay is forbidden by the conservation of baryon number.

one-third the net quark number, that is, one-third the difference between the number of quarks and the number of antiquarks. The conservation of net quark number is equivalent to the conservation of baryon number.

The skeptical reader may feel somewhat dissatisfied with baryon-number conservation as an explanation of the stability of the proton and the bound neutron. In my view he would be justified in this feeling. Baryon number was invented as a bookkeeping device, in order to explain the nonobservation of proton decays and related decays; it has no other known significance. In this respect baryon number is very different from electric charge, which

PARTICLE	MASS (MeV)	ELECTRIC CHARGE	BARYON NUMBER	PRINCIPAL DECAY MODE
(K_S^0)	497.7	0	0	$K_S^0 \rightarrow \pi^+ + \pi^-$ MASS: $497.7 \rightarrow 139.6 + 139.6$ CHARGE: $0 \rightarrow +1 + -1$ BARYON NUMBER: $0 \rightarrow 0 + 0$
(K_L^0)	497.7	0	0	$K_L^0 \rightarrow \pi^0 + \pi^0 + \pi^0$ MASS: $497.7 \rightarrow 135 + 135 + 135$ CHARGE: $0 \rightarrow 0 + 0 + 0$ BARYON NUMBER: $0 \rightarrow 0 + 0 + 0$
(K^-)	493.7	−1	0	$K^- \rightarrow \mu^- + \bar{\nu}$ MASS: $493.7 \rightarrow 105.7 + 0$ CHARGE: $-1 \rightarrow -1 + 0$ BARYON NUMBER: $0 \rightarrow 0 + 0$
PROTON (p)	938.3	+1	+1	NONE KNOWN
ANTIPROTON ($\bar{p}$)	938.3	−1	−1	NONE KNOWN
NEUTRON (n)	939.6	0	+1	$n \rightarrow p + e^- + \bar{\nu}$ MASS: $939.6 \rightarrow 938.3 + .511 + 0$ CHARGE: $0 \rightarrow +1 + -1 + 0$ BARYON NUMBER: $+1 \rightarrow +1 + 0 + 0$
ANTINEUTRON ($\bar{n}$)	939.6	0	−1	$\bar{n} \rightarrow \bar{p} + e^+ + \nu$ MASS: $939.6 \rightarrow 938.3 + .511 + 0$ CHARGE: $0 \rightarrow -1 + +1 + 0$ BARYON NUMBER: $-1 \rightarrow -1 + 0 + 0$
Λ HYPERON (Λ^0)	1115.6	0	+1	$\Lambda^0 \rightarrow p + \pi^-$ MASS: $1115.6 \rightarrow 938.3 + 139.6$ CHARGE: $0 \rightarrow +1 + -1$ BARYON NUMBER: $+1 \rightarrow +1 + 0$
Λ ANTIHYPERON ($\bar{\Lambda}^0$)	1115.6	0	−1	$\bar{\Lambda}^0 \rightarrow \bar{p} + \pi^+$ MASS: $1115.6 \rightarrow 938.3 + 139.6$ CHARGE: $0 \rightarrow -1 + +1$ BARYON NUMBER: $-1 \rightarrow -1 + 0$

Figure 53 (continued).

has a direct dynamical significance: an electric charge creates electric and magnetic fields and the charge in turn is acted on by such fields, which have observable effects on the motion of the charge. The theory of electricity and magnetism would make no sense if electric charge were not conserved, but no such dynamical argument is known for baryon-number conservation.

Indeed, there is empirical evidence against the existence of any kind of field (call it a barytropic field) that might bear the same relation to baryon number as the electromagnetic field bears to electric charge. The earth includes some 4×10^{51} protons and neutrons, and so it has a huge baryon number.

If the earth were a source of a barytropic field, one would expect the field to attract or repel the protons and neutrons in ordinary bodies on the earth's surface. A barytropic force could be distinguished from the gravitational force because the gravitational force the earth exerts on a body is proportional to the mass of the body, whereas the barytropic force would be proportional to the baryon number. Bodies of equal mass that are composed of different elements can have baryon numbers that differ by almost 1 percent. Several highly accurate experiments (starting with those of Roland von Eötvös in 1889) show that the attraction of bodies to the earth is in fact closely proportional to their mass, not to their baryon number. In 1955 T. D. Lee of Columbia University and C. N. Yang of the Institute for Advanced Study in Princeton showed from an analysis of these experiments that any barytropic force between two nuclear particles would have to be much weaker than the gravitational force, which is itself almost 40 orders of magnitude weaker than the electromagnetic force. It cannot be absolutely ruled out that baryon number plays a dynamical role like that of electric charge, but the argument of Lee and Yang makes such a role appear quite unlikely.

The conclusion that baryon number has no dynamical role does not immediately imply that baryon number is not conserved. Indeed, since the mid-1930's physicists have become familiar with a number of other quantities that do not appear to have a dynamical significance like that of electric charge and yet are conserved, at least in certain contexts. Among these quantities are the ones called strangeness, isospin and charge conjugation. To take one example, protons and neutrons are assigned a strangeness of zero, some of the hyperons are assigned a strangeness of -1 and some of the mesons called K mesons are assigned a strangeness of $+1$. The conservation of strangeness was introduced as a bookkeeping rule to explain the observation that a K meson or a hyperon cannot be produced singly in collisions of ordinary atomic nuclei but that they can be produced in association, because one K meson and one hyperon have a net strangeness of zero. For years after the idea of baryon number was introduced it did not seem implausible that baryon-number conservation was another of these nondynamical bookkeeping rules, which happens to be universally obeyed.

This view of conservation laws has been radically changed by the development of the modern theories of elementary-particle interactions. The theories describe all known forces among elementary particles (apart from gravitation) in a way very similar to the way electromagnetism was described in the older theory of purely electromagnetic interactions; the latter theory, called quantum electrodynamics, was developed in the 1930's and 1940's. There are now thought to be 12 fields similar to the electromagnetic field of quantum electrodynamics. They are the eight gluon fields that provide the strong nuclear forces that hold together the quarks inside baryons and mesons and the four electroweak fields that in a unified manner provide both

the weak nuclear forces responsible for beta decay and electromagnetism itself. There are 12 corresponding conservation laws, similar to the conservation of electric charge, for quantities designated color, electroweak isospin and electroweak hypercharge. (Color is a property of quarks that has nothing to do with visual color; electric charge is a particular weighted combination of electroweak hypercharge and electroweak isospin.) Unlike baryon number, these conserved quantities have a direct physical significance: it is the particles carrying these quantities that give rise to the gluon fields and the electroweak fields, and the fields in turn exert a force on any such particle. The force depends on the values of the 12 quantities carried by the particle.

At the same time that new conservation laws have appeared the old nondynamical conservation laws have in a sense been demoted. For example, the modern theory of strong nuclear interactions is so tightly constrained by color conservation (and other principles) that there is no way it could include the kinds of complications that would be needed to violate the conservation of strangeness. One could try to introduce fundamental interactions that do not conserve strangeness, but it always turns out that one can redefine what is meant by strangeness in such a way that it is still conserved. Thus strangeness conservation is now understood to be not a fundamental principle like energy conservation or charge conservation but a consequence of the detailed theory of the strong interactions, and in particular of the genuinely fundamental law of color conservation. Since strangeness conservation is not a fundamental principle of physics, there is no general reason for it to be respected outside the realm of the strong forces. Indeed, it has been known since the discovery of strangeness that strangeness is not conserved by the weak nuclear forces.

The other nondynamical conservation laws have suffered a similar demotion in status; they are no longer seen as fundamental conservation laws, on the level of energy or charge conservation, but rather as mere mathematical consequences of the structure of present theories of elementary-particle interactions. A list of the conservation laws that now seem to be fundamental would include the conservation of the 12 quantities associated with the strong and the electroweak forces, the conservation of quantities such as energy and momentum that are associated in a similar way with gravitational forces, and the conservation of baryon number, which is not known to be associated with any force.

This fact alone should make one suspicious about baryon-number conservation: baryon number does not need to be conserved in the way that energy, charge, color and similar quantities need to be conserved in order to have sensible theories of elementary-particle interactions. Moreover, there are positive hints that baryon-number conservation is not exact. One of the hints is provided by the modern theory of electroweak interactions. Gerard 't Hooft of the University of Utrecht has shown that in this theory certain subtle effects that cannot be represented by any finite number of emissions and absorptions of elementary particles lead to baryon-nonconserving processes, but they are processes with an extraordinarily low rate. The processes are much too slow to be detected, but it is interesting that they arise precisely because baryon-number conservation is not related to any kind of barytropic field; no such effects could produce a nonconservation of quantities, such as electric charge, that are related to fields of force.

Another hint that baryon number may not be conserved comes from cosmology. One might have supposed, if only on aesthetic grounds, that the universe began with equal amounts of matter and antimatter and therefore with equal numbers of baryons and antibaryons. On this hypothesis the universe would have started with a total baryon number equal to zero. If baryon number were conserved, the total baryon number would have remained zero. Almost all protons and neutrons would have been annihilated through collisions with antiprotons and antineutrons and the universe today would contain only a thin gruel of photons and neutrinos, with no stars or planets or scientists.

It is possible the universe started with an excess of matter over antimatter, so that something would be left over after the annihilation of particles and antiparticles. It is also possible (although it is generally regarded as unlikely) that matter and antimatter have somehow become segregated and that we live in a patch of positive baryon number in a universe with a total baryon number of zero. If baryon number is not conserved, however, there is a more appealing possibility, namely that the universe did start with equal amounts of matter and antimatter and that the present excess of particles with positive baryon number is due to physical processes that

have violated the conservation of baryon number. (It has been known since the 1964 experiment of James H. Christenson, James W. Cronin, Val L. Fitch and René Turlay of Princeton University that there is no exact matter-antimatter symmetry that would require processes in which antibaryons are created to go on at the same rate as those that create baryons.) These considerations, together with the absence of barytropic forces as shown by the arguments of Lee and Yang, led some theorists (including the Russian physicist Andrei D. Sakharov and me) to suggest in the 1960's that baryon number may not be exactly conserved. Cosmological considerations also instigated at least one of the proton-decay experiments done in this period, that of T. Alväger, I. Martinson and H. Ryde of the University of Stockholm and the Nobel Institute. In recent years many theorists have worked out schemes for the production of baryons in the very early universe.

Any suggestion of possible baryon-number nonconservation has to immediately confront the fact that ordinary matter is very stable. Maurice Goldhaber of the Brookhaven National Laboratory has remarked that "we know in our bones" the average lifetime of the proton is longer than about 10^{16} years. If the lifetime were any shorter, the 10^{28} or so protons in the human body would be decaying at an average rate of more than 10^{12} protons per year, or 30,000 decays per second, and we would be a health hazard to ourselves.

Of course, one can set a more stringent limit on the proton lifetime by actively searching for proton decay. The first experiment of this kind was carried out in 1954 by Frederick Reines and Clyde L. Cowan, Jr., who were then at the Los Alamos Scientific Laboratory, and Goldhaber. They used about 300 liters of hydrocarbon scintillator, a material in which the energetic charged particles produced by a proton decay would generate a detectable flash of light. As in all subsequent proton-decay experiments, the apparatus was placed underground to shield it from cosmic rays. (The energetic particles of these rays can cause events that might be mistaken for proton decay.) With this precaution they observed only a few scintillations per second, almost all of which could be attributed to cosmic rays that penetrated deep underground. Reines, Cowan and Goldhaber concluded that the average lifetime of a proton or of a bound neutron must be greater than about 10^{22} years.

Subsequent experiments done by a number of physicists have gradually increased the empirical lower bound on the lifetime of the proton. The most elaborate search so far whose results have been published was undertaken by a consortium of investigators from Case Western Reserve University, the University of the Witwatersrand and the University of California at Irvine. They monitored 20 tons of a hydrocarbon scintillator at a depth of 3.2 kilometers in a South African gold mine from 1964 to 1971. A recent analysis of their data gave the result that the average lifetime of a proton or a bound neutron is longer than about 10^{30} years.

That is truly a long lifetime. For the purpose of comparison the present age of the universe is estimated to be a mere 10^{10} years. One can hope to observe the decay of particles with such long lifetimes only because radioactive-decay processes operate statistically: a sample of particles with an average lifetime of t years will not all survive for t years and then decay in unison; rather a fraction $1/t$ of the total number of particles will decay in the first year, $1/t$ of the remaining particles will decay in the next year, and so on. The lower bound on the proton lifetime is set not by watching one proton for a long time and waiting for it to decay but by watching the 10^{31} protons and neutrons in 20 tons of scintillator for several years and waiting for a few dozen to decay.

It is the long lifetime of the proton that led to the idea of baryon-number conservation. How could the proton live so long if there were not some conservation law that makes it live forever? In the past few years an answer has emerged.

Remember that the modern theory of the weak, the electromagnetic and the strong interactions is highly constrained, so much so, for example, that it is impossible for the strong interactions to violate the conservation of the quantity called strangeness. It happens the theory is so constrained that it cannot be complicated enough (apart from the tiny 't Hooft effect) to allow for any violation of baryon-number conservation unless one introduces new kinds of particles with exotic values of charge, color and so on. Such particles would have to be qualitatively different from any particles now known.

If exotic particles of the right kind are admitted, proton decay becomes a possibility. The familiar conservation laws for charge, color and so on indicate that what is needed is a particle with an electric charge of $+4/3$, $+1/3$ or $-2/3$ times the charge of the proton; the particles must also have an intrinsic spin

angular momentum equal to 0 or 1 and a color identical with that of the antiquark. For example, such an exotic particle could be produced when a quark turns into an antiquark and could then be destroyed when another quark turns into an antilepton (a positron, an antimuon or an antineutrino); in this way the three quarks that make up the proton could decay into an antilepton and a meson formed from the leftover quark and the antiquark.

Any such exotic particles would have to be very heavy or they would already have been detected. If they were heavy enough, they would be emitted and reabsorbed only with difficulty and hence would induce only a very small rate of proton decay. Thus it is now possible to explain the long lifetime of the proton without assuming any fundamental conservation law that would ensure it lives forever, and this has opened up the possibility that it does not live forever.

How heavy would the exotic particles have to be in order to explain the long lifetime of the proton? Assuming that the exotic particles interact more or less like photons, one can roughly estimate that a proton lifetime longer than 10^{30} years requires an exotic particle whose mass is greater than about 10^{14} proton masses. This is a stunningly large mass, larger than anyone can hope to produce with any accelerator that can now be envisioned. Yet there are at least two reasons for suspecting such enormously heavy particles may actually exist.

The first reason has to do with the phenomenon of gravitation, which has so far been left out of these considerations. Einstein's general theory of relativity provides a satisfactory theory of gravitational interactions of particles at all experimentally accessible energies. Because of quantum fluctuations, however, the theory breaks down at very high energies, on the order of 10^{19} proton masses. This is known as the Planck mass, after Max Planck, who noted in 1900 that some such mass would appear naturally in any attempt to combine his quantum theory with the theory of gravitation. The Planck mass is roughly the energy at which the gravitational force between particles becomes stronger than the electroweak or the strong forces. In order to avoid an inconsistency between quantum mechanics and general relativity, some new features must enter physics at some energy at or below 10^{19} proton masses.

The other reason for expecting new degrees of freedom to appear at superlarge energies has to do with the electroweak and the strong interactions.

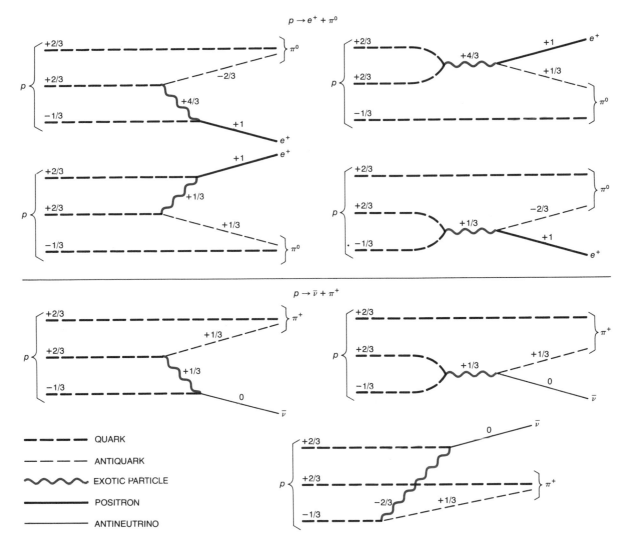

Figure 54 DECAY OF THE PROTON by the emission and absorption of a heavy exotic particle has the effect of transmuting two quarks into an antiquark and a positron (*upper four reactions*) or into an antiquark and an antineutrino (*lower three reactions*). Exotic particles are distinguished by their electric charge (+ or −⅓, + or −⅔ and + or −⅘) and by their intrinsic spin angular momentum (either equal to zero or 1).

The modern theory of these interactions involves three parameters, which are known as coupling constants. One of the coupling constants, called g_s, describes the strength with which the gluon fields of the strong force interact with particles that carry the conserved quantities designated color; the other two coupling constants, called g_1 and g_2, describe the strength with which the electroweak fields interact with particles carrying the corresponding quantities, electroweak hypercharge and electroweak isospin. One would like to believe all these interactions have some common origin, in which case the coupling constants should all have the same order of magnitude. But this is in apparent disagreement with the obvious fact that the strong interactions are strong; measurements yield a value for g_s that is much larger than the value of g_1 or g_2. (See Chapter 4, Figure 37.)

A solution to this difficulty was proposed at Harvard University in 1974 by Howard Georgi, Helen R. Quinn and me. It has been known since the 1954 work of Murray Gell-Mann of the California Institute of Technology and Francis E. Low of the Massachusetts Institute of Technology that coupling constants depend somewhat on the energy of the physical processes in which they are measured. In 1973 independent calculations by H. David Politzer of Harvard and David Gross and Frank Wilczek of Princeton showed that the strong coupling constant g_s decreases slowly with increasing energy. The larger of the two electroweak coupling constants, g_2, also decreases, but more slowly, whereas the smaller electroweak coupling constant, g_1 increases with increasing energy. What Georgi, Quinn and I proposed was that the scale of energies at which the strong interactions become unified with the electroweak interactions is enormously high, so high that the very slow decease with increasing energy of the strong coupling constant and the even slower variation with energy of the two electroweak coupling constants bring them all to essentially the same value at this superhigh energy scale. Specifically, under rather general assumptions (in essence, that the strong and the electroweak interactions are unified by some set of symmetries of the kind known mathematically as a "simple" group, that there are no intermediate stages of unification and that elementary particles with one-half unit of spin form patterns more or less like the familiar patterns of the leptons and the quarks) we found that the energy scale at which the strong and the electroweak interactions are unified is on the order of 10^{15} to 10^{16} proton masses.

Any theory that unifies the strong and the electroweak interactions and puts leptons and quarks on the same footing would have to involve new kinds of particles to fill out the picture, and, as I have argued above, there is no good reason to think the interactions of those new particles would conserve baryon number. (The energy scale of 10^{15} to 10^{16} proton masses calculated by Georgi, Quinn and me is sufficiently high so that baryon-number nonconserving interactions that could be produced by exotic particles this heavy would not lead to a proton lifetime in contradiction with the present experimental lower limit of 10^{30} years. We estimated a lifetime very roughly on the order of 10^{32} years.)

Starting in 1973 many theorists have worked to develop such theories, including Jogesh C. Pati of the University of Maryland and Abdus Salam of the International Centre for Theoretical Physics in Trieste, Georgi and Sheldon Lee Glashow of Harvard, Harold Fitsch and Peter Minkowski of Cal Tech and Feza Gürsey, Pierre Ramond and Pierre Sikivie of Yale University. The models are generally known by the name of the mathematical group of symmetries that connects the various forces, such as $SU(4)^4$, $SU(5)$, $SO(10)$, E_6, E_7 and $SU(7)$. These models all include exotic particles that when emitted or absorbed convert a quark into an antiquark, a lepton or an antilepton; therefore, as already pointed out by Pati and Salam in the first paper on the unification of the strong and the electroweak interactions, they can violate the conservation of baryon number. Furthermore, all these models in at least some of their versions satisfy the general assumptions made by Georgi, Quinn and me, so that the mass scale of the exotic particles would be expected to be on the order of 10^{15} to 10^{16} proton masses and the proton lifetime would be on the order of our rough estimate, 10^{32} years.

More recent refined calculations have been made by many theorists, including Andrzej Buras, John Ellis, Mary K. Gaillard and Demetres V. Nanopoulos of the European Organization for Nuclear Research (CERN), Terrence J. Goldman and Douglas A. Ross of Cal Tech, William J. Marciano and Alberto Sirlin of Rockefeller University and New York University, Cecilia Jarlskog and Francesco Yndurian of CERN and Lawrence Hall of Harvard. The newer calculations give an improved value of about 10^{15} proton masses for the superheavy-mass scale and a proton lifetime of about 10^{31} years. Unfortunately the calculation of the proton-decay rate is complicated by the presence of strong nuclear forces acting among the quarks and antiquarks in the proton and in the decay products, and so even if the properties of the superheavy exotic particles were known precisely, it would probably remain impossible to predict the proton lifetime more accurately than to within an order of magnitude or so.

Experimental studies of the weak interactions have already provided limited verification of the general analysis made by Georgi, Quinn and me. It would be no surprise to find that the graphs for any two of the three coupling constants intersect somewhere, but in order for the three curves of coupling constant v. energy to cross at the same point it is necessary to impose one condition on their starting points, that is, on the values of the coupling constants at low energies. We used this condition to calculate that under our general assumptions a certain parameter (related to the ratio of g_1 and g_2), which describes the unification of the weak and the electromagnetic interactions, has a value close to .2. Experiments in the physics of electron and neutrino interactions currently indicate a value for this parameter of from about .2 to .23. The theoretical and experimental values are close enough to encourage us to take this analysis seriously even though it involves an extrapolation of an unprecedented extent: by 13 orders of magnitude in energy.

A mass on the order of 10^{15} proton masses is so large that the emission and absorption of particles this heavy is almost impossible at experimentally accessible energies and can therefore produce only tiny effects in any feasible experiment. The only hope for detecting these tiny effects lies in the possibility that they may violate otherwise exact conservation laws and thereby make possible processes that otherwise would be strictly forbidden. One of these conservation laws is baryon-number conservation, which is tested by looking for proton decay. The only other known conservation law that is not required for the consistency of theories of particle interactions and therefore could be violated by superhigh-energy effects is lepton-number conservation, the conservation of the total number of leptons (neutrinos, electrons, muons and so on) minus the number of antileptons. The nonconservation of lepton number could show up in such processes as neutrinoless double beta decay: the decay of two protons in a nucleus into two neutrons plus two positrons. Baryon number would remain constant in this reaction, but lepton number would decrease by 2. (The reaction does not violate the conservation of energy because the neutrons in the final state are not free but have a negative binding energy in the nucleus.) Another sign of the violation of lepton-number conservation would be a nonzero neutrino mass.

Several new attempts to determine the lifetime of the proton have now been carried out. Experiments are already in operation in the Soudan mine in Minnesota, the Kolar gold field in southern India, the Mont Blanc tunnel between France and Italy and the Baksan valley in the Caucasus range of the U.S.S.R. Other experiments include the Morton salt mine in Ohio, the Silver King mine in Utah and other sites in the Mont Blanc tunnel and in Kamioka, Japan.

The basic technique of all the experiments is to compensate for the extreme slowness of proton decay by careful monitoring of a very large mass of material. The larger the mass, the larger the number of protons and bound neutrons and hence the greater the probability of observing a decay. In this way it is expected that it will be possible to detect the decay of the proton or the bound neutron with an average lifetime rather longer than the current bound of 10^{30} years. The experiments differ chiefly in the nature and amount of material monitored, the nature and arrangement of the devices used to detect a proton decaying in the material and the characteristics of the experiment that suppress spurious signals from cosmic rays, including the depth underground at which the experiments are carried out.

Since a very large mass must be monitored, the experiments must use some relatively inexpensive material, such as water, concrete or iron. In experiments such as those in the Soudan mine, the Kolar gold field and the Mont Blanc tunnel, which use iron or concrete, one must rely on detection devices such as proportional tubes or streamer tubes, which can directly detect the energetic charged particles that are expected to be emitted in proton decays. The charged particles have a short range in iron or concrete, and so the detector tubes must be closely spaced throughout the monitored material.

On the other hand, experiments such as those in the Morton salt mine, the Silver King mine and the Homestake gold mine, which monitor a transparent material such as water, can use a rather different strategy. The energy released in proton decay is great enough so that an electron, a positron, a muon or a pi meson emitted in the decay process is likely to have a very high speed, slower of course than the speed of light in vacuum but faster than the speed of light in water. When a charged particle travels through a transparent medium at a speed higher than the speed of light in that medium, there occurs what is known as the Cerenkov effect (see Chapter 7, Figure 58). It is like the sonic boom generated by an airplane traveling faster than the speed of sound in air, but the Cerenkov effect is an optical boom, in which the particle emits a cone of light rather than of sound. (The beautiful blue glow of Cerenkov light was noted in early experiments on radioactivity by Marie Curie, but its properties were first explored in detail in the 1930's by Pavel A. Cerenkov.) The angle between the Cerenkov light rays and the path of the charged particle depends on the ratio of the speed of the charged particle to the speed of

Figure 55 LARGE CAVITY some 1,950 feet underground in the Morton salt mine east of Cleveland is filled with water to search for the decay of a proton or a neutron bound in an atomic nucleus. The decay of any one of the 2.5×10^{33} protons and neutrons in the central region of the water gives rise to high-velocity particles that are detected by photomultiplier tubes installed along the walls of the cavity.

light in the medium. Most of the light is emitted by particles traveling at nearly the speed of light in vacuum, and in this case the angle has a characteristic value in water of approximately 42 degrees.

Observation of a cone of Cerenkov light is a signal that something has happened in the medium to create a fast charged particle. Furthermore, for a given initial speed, the thickness of the cone and the amount of light emitted depend only on the distance the charged particle travels before its speed drops below the speed of light in the medium, which in turn depends on its initial energy. By registering the positions where light is received and the light's intensity, one can therefore deduce the initial energy as well as the direction of each charged particle. In certain circumstances the moving particle itself might decay, emitting other particles that could give rise to a second flash of Cerenkov light. A muon or an antimuon created in a proton decay could decay into an electron that emitted Cerenkov radiation. A charged pi meson might decay into a slow muon (or antimuon), which in turn could decay into a fast electron (or positron) that would give off a cone of light. A neutral pi meson could decay into two photons, each of which would give rise to a shower of charged particles and the accompanying Cerenkov light. The use of Cerenkov light to detect proton decays thus offers an alternative to the direct observation of charged particles as a means of reconstructing the decay process and verifying that one is really seeing a proton decay.

One advantage of using Cerenkov light over other means of detecting proton decay is that the light can travel greater distances in water than the charged particles themselves can. Hence with a given volume of monitored material one needs fewer detectors than one does in experiments that use opaque materials such as iron or concrete. Also, water is cheaper than iron or concrete. On the other hand, water Cerenkov detectors are sensitive only to charged particles traveling faster than the speed of light in water. Also, the relatively low density of water requires that larger cavities be excavated underground to hold a given mass of monitored material, and the water has to be kept very pure to maintain its transparency to Cerenkov light. As I have indicated, work is proceeding vigorously both on experiments using water as the monitored material and on those using denser materials.

What proton-decay rates could be detected in these experiments? To take one example, the experiment with the largest monitored mass is the one in the Morton salt mine. Of its monitored mass of 10,000 tones of water, an outer shell of perhaps 5,000 tons is used to keep track of the background of cosmic rays. The remaining mass of about 5,000 tons of water include 3×10^{33} protons (and bound neutrons). If the proton has the average lifetime of about 10^{31} years that is indicated by refined versions of the analysis done by Georgi, Quinn and me there should be some 300 proton-decay events per year. A few years of observation would yield a few proton-decay events even if the lifetime were as long as 10^{33} years, but at such low decay rates the experiments would start to be jeopardized by an unremovable background of spurious events due to cosmic-ray neutrinos, and further improvement would be difficult.

What will be learned if proton decay is discovered? Of course one will immediately conclude that baryon number is not conserved, and this will support the growing belief that all conserved quantities have a dynamical significance similar to that of electric charge. Furthermore, if proton decay is discovered in the near future, the lifetime will have to be in the range of from 10^{30} to 10^{33} years, and this will lend some credence to the general assumptions about the unification of the strong and the electroweak forces that were used by Georgi, Quinn and me. There are a great many theories, however, that satisfy these general assumptions, including at least some versions of the models $SU(4)^4$, $SU(5)$, $SO(10)$ and so on mentioned above. It will be difficult to tell which (if any) of these theories describes physics at very high energy.

About one thing we can be sure. If proton decay is discovered, great new resources will be devoted to its study, and before long there will be a second generation of experiments, in which the effort will be to learn not whether protons decay but how they decay: what are the probabilities for the various modes of decay?

In preparation for this effort a number of theorists have been exploring the likely modes of proton decay. (The remarks below are based on independent work by Wilczek and Anthony Zee and by me.) Interestingly, one can go pretty far in this analysis without any assumptions about the unification of the strong and the electroweak interactions. All that one needs are the familiar conservation laws for charge, color and so on and the assumption that the exotic particles responsible for proton decay are very heavy, as they surely must be to explain the long proton lifetime. In this case, although the emission and reabsorption of these particles can

LOCATION	DATE OF OPERATION	DEPTH (EQUIVALENT METERS OF WATER)	MONITORED MATERIAL
HOMESTAKE GOLD MINE, SOUTH DAKOTA	NOW	4,400	150 TONS OF WATER (TO BE INCREASED TO 900 TONS)
KOLAR GOLD FIELD, INDIA	NOW	7,600	150 TONS OF IRON (½-INCH PLATES)
BAKSAN VALLEY, U.S.S.R.	NOW	850	80 TONS OF LIQUID SCINTILLATOR
SOUDAN MINE, MINNESOTA	NOW	1,800	30 TONS OF TACONITE CONCRETE AND IRON (TO BE INCREASED TO 1,000 TONS)
MONT BLANC TUNNEL BETWEEN FRANCE AND ITALY	NOW	4,270	30 TONS OF IRON AND LIQUID SCINTILLATOR (TO BE INCREASED TO 200 TONS)
MORTON SALT MINE, OHIO	1981 ESTIMATED	1,670	10,000 TONS OF WATER
SILVER KING MINE, UTAH	1981 ESTIMATED	1,700	1,000 TONS OF WATER
MONT BLANC TUNNEL BETWEEN FRANCE AND ITALY	1981 ESTIMATED	5,000	150 TONS OF IRON INITIALLY (ONE-CENTIMETER-THICK PLATES)
GRAND SASSO TUNNEL, ITALY	UNDER CONSTRUCTION	4,000	10,000 TONS OF IRON (THREE-MILLIMETER PLATES)
JAPAN	UNDER CONSIDERATION	2,700	3,400 TONS OF WATER
FRÉJUS TUNNEL BETWEEN FRANCE AND ITALY	UNDER CONSIDERATION	4,500	1,500 TONS OF IRON (PLATES THREE TO FOUR MILLIMETERS THICK)
JAPAN	UNDER CONSIDERATION	2,700	600 TONS OF IRON
ARTYEMOVSK SALT MINE, U.S.S.R.	UNDER CONSIDERATION	600	100 TONS OF LIQUID SCINTILLATOR

**Figure 56 TABLE OF PROTON-DECAY EXPERIMENTS.
The experiments differ in the kind and amount of material
monitored. Moreover, they are done at different depths in
the ground. The depth is given in terms of the equivalent
depth of water that would provide the same shielding from
cosmic rays.**

produce a great many different modes of proton
decay, the more complicated modes are more
strongly suppressed by the large mass of the exotic
particles than the simpler modes. Unless some special circumstance intervenes, the dominant modes
of decay will in general be those in which the proton or the bound neutron decays into a positron, an
antimuon or an antineutrino, plus some number of
mesons. Once can go further and make predictions
about ratios of decay rates. For instance, a neutron
decays just twice as fast as a proton into a positron
and a single pi meson or rho meson. The proton
decays faster into a positron plus mesons than a

neutron decays into an antineutrino plus mesons.
The neutron decays faster into a positron and
mesons than the proton decays into an antineutrino
and mesons.

One cannot be sure these predictions will be
borne out by experiment. If they are not, there
must be exotic particles considerably lighter than
10^{14} proton masses to produce more complicated
decay modes. For example, the decay of a proton or
a bound neutron into an electron and mesons rather
than into a positron and mesons could be produced

DETECTORS	INSTITUTIONS
144 PHOTOMULTIPLIER TUBES IN WATER	UNIVERSITY OF PENNSYLVANIA
GAS PROPORTIONAL COUNTERS BETWEEN PLATES	TATA INSTITUTE OF FUNDAMENTAL RESEARCH, BOMBAY UNIVERSITY OF OSAKA UNIVERSITY OF TOKYO
1,200 PHOTOTUBES	INSTITUTE FOR NUCLEAR RESEARCH, MOSCOW
3,456 GAS PROPORTIONAL COUNTERS IN CONCRETE	UNIVERSITY OF MINNESOTA ARGONNE NATIONAL LABORATORY
PHOTOTUBES AND STREAMER CHAMBERS	INSTITUTE FOR NUCLEAR RESEARCH, MOSCOW UNIVERSITY OF TURIN
2,400 PHOTOMULTIPLIER TUBES IN WATER	UNIVERSITY OF CALIFORNIA AT IRVINE UNIVERSITY OF MICHIGAN BROOKHAVEN NATIONAL LABORATORY
800 PHOTOMULTIPLIER TUBES IN WATER AND MIRRORED WALLS	HARVARD UNIVERSITY PURDUE UNIVERSITY UNIVERSITY OF WISCONSIN
STREAMER CHAMBERS BETWEEN PLATES	ITALIAN NATIONAL SYNCHROTRON LABORATORY AT FRASCATI UNIVERSITY OF MILAN UNIVERSITY OF TURIN
FLASH CHAMBERS TRIGGERED BY STREAMER CHAMBERS	ITALIAN NATIONAL SYNCHROTRON LABORATORY AT FRASCATI UNIVERSITY OF MILAN UNIVERSITY OF TURIN UNIVERSITY OF ROME
1,056 20-INCH PHOTOMULTIPLIER TUBES	UNIVERSITY OF TOKYO JAPANESE NATIONAL LABORATORY FOR HIGH-ENERGY PHYSICS UNIVERSITY OF TSUKUBA
FLASHTUBES AND STREAMER CHAMBERS	ÉCOLE POLYTECHNIQUE UNIVERSITY OF PARIS AT ORSAY SCHOOL OF MINES AT SACLAY
NEON FLASHTUBES	UNIVERSITY OF OSAKA UNIVERSITY OF TOKYO JAPANESE NATIONAL LABORATORY FOR HIGH-ENERGY PHYSICS
128 PHOTOMULTIPLIER TUBES	INSTITUTE FOR NUCLEAR RESEARCH, MOSCOW

at an observable rate if these were exotic particles no heavier than about 10^{10} proton masses. A decay into three neutrinos or three electrons (or some other combination of three leptons) could be observable if there were exotic particles no heavier than about 10^4 proton masses. Such relatively light exotic particles, however, would have to have special properties to avoid producing the "ordinary" decay of the proton (into a meson and a positron or an antineutrino) at much too high a rate.

The verification of these predictions, that is, the finding that the proton decays into mesons and a positron or an antineutrino and with the above ratios of decay rates, would serve to confirm that proton decay is really due to exotic particles with masses greater than about 10^{10} proton masses, but it would not point toward any specific underlying theory. For that purpose it would be necessary to explore finer details of the decay process. (For example, a determination of the direction in which a positron or an antimuon spins when it is produced in a proton decay can be used to diagnose the spin of the superheavy exotic particles whose emission and reabsorption produced the decay.) If proton decay is discovered, it will rank as a triumph of experimental ingenuity and an unparalleled clue to the physics of very high energies, but it will also present experimental and theoretical physicists with many new tasks that will need to be done if the mechanism of proton decay is to be understood.

The Search for Proton Decay

Physicists have been keeping watch over an 8,000-ton underground detector, waiting for a sign that all matter has a finite lifetime. So far no proton has been observed to decay, but the vigil will continue.

. . .

J. M. LoSecco, Frederick Reines and Daniel Sinclair

Life is fleeting and kingdoms fall, and even stars and galaxies may someday fade, yet one might think the basic stuff of matter — the protons, neutrons and electrons of the atom — would endure forever. In the case of the electron it is probably so: both experiment and the elegance of physical theory suggest that electrons are immune to decay. For the proton and the neutron, however, immortality is far from certain. Indeed, the neutron, when it is not stably bound inside the nucleus of an atom, is known to decay. It breaks down spontaneously to yield three lighter particles: a proton, an electron and a neutrino. What keeps the proton from decaying into lighter particles? There has never been a secure basis for the belief that it does not. A proton might break down, for example, into a positron (a positively charged electron) and a pair of neutrinos, and no general or fundamental law of physics would be violated. It seems there is nothing in nature to prevent this process; on the other hand, it has never been observed.

The possibility of proton decay has been a nagging question in physics since the 1930's , but in the past 10 years it has become a matter of more than passing interest. A new class of theories has been developed in which the decay of the proton is not only allowed but also definitely predicted. With the simplest of the theories it is even possible to calculate the proton's lifetime: the estimate is about 10^{30} years, many orders of magnitude greater than the age of the universe (roughly 10^{10} years), but finite all the same. Observation of the decay and measurement of the lifetime are the chief means of testing the theories.

Experimental physicists have taken up the challenge. The methods needed for the search are quite different from those of most experiments in elementary-particle physics. Instead of bombarding a detector with a beam of high-energy particles, every effort is made to shield the apparatus from stray particles that might strike it. Accordingly, enormous detectors have been set up deep underground in tunnels and mine shafts. No one has yet observed an unequivocal instance of proton decay, and as a result the experimental lower limit on the proton's lifetime is now greater than the theoretically predicted lifetime. This does not mean that the proton is stable, however; it means the search for proton decay must go on. (See Figure 57, and Chapter 6, Figure 54.)

Figure 57 SIMULATED DECAY OF A PROTON on a graphic display of a large detector. When light from the particles strikes a photomultiplier tube, the tube "fires," producing an electrical pulse, marked by a star, and the number of points in the star indicates the amount of light received. Color indicates timing; tubes marked by a red star are first to fire, followed by yellow, green and blue. Reconstructed back-to-back tracks of the decay products are indicated by heavy yellow lines.

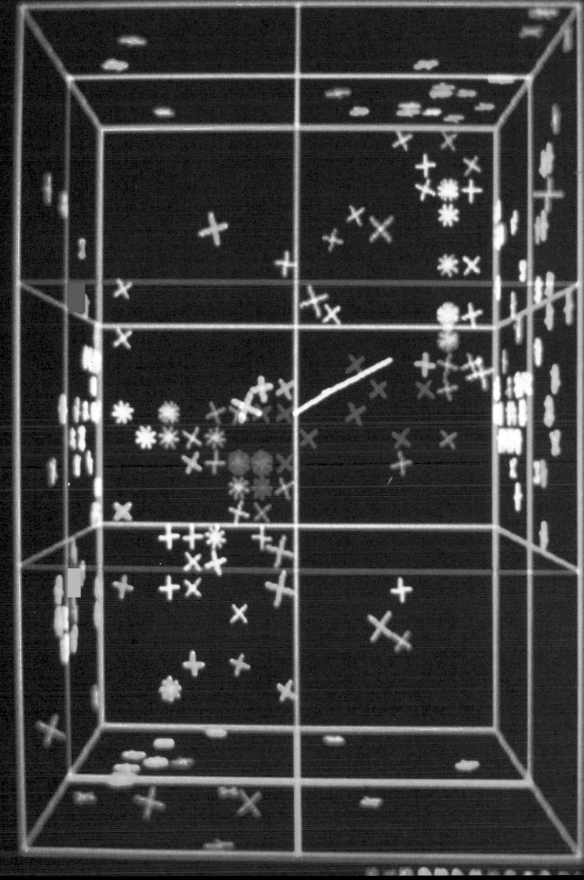

The rules governing the decay of particles are conservation laws, which state that a certain property or quantity must remain forever unchanged. The most important laws for our purposes require the conservation of energy, linear momentum, angular momentum and electric charge. In general any particle will decay unless it is prevented from doing so by one of these laws. The decay of an isolated neutron described above is consistent with all four laws; for example, the total electric charge is zero both before and after the decay. The electron, on the other hand, cannot decay because it is the lightest particle with an electric charge; any imagined scheme of electron decay would violate the law of charge conservation.

It should be emphasized that the four conservation principles cited here are quite general and are grounded in fundamental concepts. They are useful in all domains of physics, and their validity has been proved. The puzzling thing about the proton's apparent stability is that no such general law accounts for it. The hypothetical decay of a proton into a positron and two neutrinos would conserve energy, linear momentum, angular momentum and electric charge. Many other possible decay modes would also obey the conservation rules.

In the 1930's Hermann Weyl and E. C. G. Stückelberg, and later Eugene P. Wigner, attempted to explain the proton's stability by postulating a new conservation law. They put the proton and the neutron in a class of particles called baryons and assigned a "baryon number" of +1 to them; they then proposed that baryon number is a conserved quantity in nature. The neutron can decay into a proton, an electron and a neutrino because the total baryon number is unchanged by the process (it is +1 both before and after the decay). The proton cannot decay because it is the lightest baryon.

The introduction of baryon number does not really explain what inhibits the proton's decay but merely gives it a name. The proposed law of baryon conservation is not a general law of physics: it has no application beyond the field of elementary particles, and it is not founded on any fundamental concept. Furthermore, the law would be suspect even if it were not arbitrary. The universe is composed almost entirely of protons and electrons. For each positive electric charge (proton) there is evidently a negative charge (electron), so that the universe as a whole is electrically neutral. It is not neutral with respect to baryon number, however. There are far more protons (baryon number +1)

than antiprotons (baryon number −1). It appears that in the extremely hot first moments of the big bang more protons than antiprotons were created, and so some process operating then must have violated the law of baryon conservation. If the law could be broken then, why not now?

Although the conceptual underpinnings of proton stability have always been shaky, the problem was given little attention until 1974, when new theoretical models called grand unified theories suddenly made it an issue of immediate concern. The goal of the new theories is to unify three of the four fundamental forces of nature. The strong and the weak nuclear forces and the force of electromagnetism are to be made part of a single framework, leaving only gravitation as a separate entity. There are precedents in physics for such a unification. In the 19th century James Clerk Maxwell unified the theories of electricity and magnetism, and in the 1960's a deep connection was found between the weak force and electromagnetism.

The simplest of the grand unified theories was developed by Howard Georgi and Sheldon Lee Glashow of Harvard University; it is called minimal $SU(5)$. The designation $SU(5)$ refers to the mathematical group of symmetries on which the theory is based; it is minimal in that it is the theory with the fewest "adjustable parameters," which must be assigned a value by experiment. According to minimal $SU(5)$, the strong, weak and electromagnetic forces, which seem very different under ordinary circumstances, become indistinguishable when particles interact with an energy of approximately 10^{15} billion electron volts (GeV). Moreover, at this enormous energy the law of baryon conservation is repealed: events that change baryon number can take place as readily as events that conserve it.

The unification energy of 10^{15} GeV is far beyond the reach of laboratory experiments: the largest particle accelerators have yet to surpass 1,000 GeV. The failure of baryon conservation has consequences even for matter at rest, however. In particular, minimal $SU(5)$ predicts that a proton can decay by means of an intermediate state with an energy, or mass, of 10^{15} GeV; the intermediate state decays in turn to yield light particles. It might seem this process would violate the conservation of energy, since a proton with a mass of less than 1 GeV gives rise to an intermediate particle of much greater mass. The intermediate state is so short-lived, however, that it cannot be detected even in principle; from the point of view of observation it does not exist, and energy is conserved.

Even though the great mass of the intermediate state does not forbid proton decay, it does greatly diminish the probability of the event. At any given instant a proton is most unlikely to emit a particle with a mass of 10^{15} GeV. Because the decay process is highly improbable, the lifetime of the proton is extremely long. In minimal $SU(5)$ the estimated lifetime is approximately 10^{30} years. Other grand unified theories also predict the decay of the proton, but the theories are too complicated to support a calculation of the lifetime.

Although we have been speaking exclusively of the proton's fate, the neutron is subject to decay by the same mechanism. An isolated neutron can decay (as outlined above) to yield a proton, an electron and a neutrino, but when the neutron is bound in an atomic nucleus, the process is suppressed. (The reason is that adding the positive charge of a proton to a nucleus that already has a positive charge costs more energy than is released in the decay.) Thus a bound neutron is stable against all modes of decay that conserve baryon number, but it can decay if baryon conservation is violated. In the grand unified theories the lifetime of a bound neutron is similar to the lifetime of a proton, but the decay modes would be different because of the difference in electric charge.

How can one measure a lifetime of 10^{30} years in a universe that is only 10^{10} years old? There is no need to wait for a particular, selected proton to decay. The estimate of 10^{30} years represents the half-life of the proton: the time in which half of the protons in any sample of matter can be expected to decay. With a large enough sample some events will be seen even in a much shorter period; monitoring 10^{30} protons ought to yield an average of one decay per year.

The most straightforward approach to detecting the decay of the proton would be an experiment based on simple counting. All the protons in a large sample of matter would be counted and then the matter would be set aside for a year or so. If a second count revealed that some protons were missing, they could be assumed to have decayed. Such an experiment would be independent of all assumptions about the mode of the proton's decay: the set of particles emitted is immaterial, since what is detected is the absence of the proton itself. Unfortunately the experiment is totally impractical. It would require counting 10^{30} protons without error.

Another approach is not only practical but also easy. Proton decay constitutes a kind of radioactivity, and so it must contribute to the background radiation at the earth's surface. The background can be measured with a Geiger counter or a similar instrument; if the contributions from known sources were then subtracted, one could assume that whatever is left is due to proton decay. This crude procedure sets a lower limit on the proton lifetime of 10^{17} years, or 10 million times the age of the universe.

Until the development of the grand unified theories the only incentive for mounting more elaborate experiments was the conviction that a principle, such as baryon conservation, is no better than the experiments that test it. In accordance with this precept a measurement of the proton lifetime was undertaken in 1953 by Clyde L. Cowan, Jr., of the Los Alamos Scientific Laboratory, Maurice Goldhaber of the Brookhaven National Laboratory and one of us (Reines). It was the first experiment to employ a massive detector to search for proton decay.

The detector was a 300-liter tank filled with a liquid scintillator, a substance that emits a flash of light when a charged particle passes through it. Ninety photomultiplier tubes registered the flashes. The detector had originally been designed for another purpose (experiments with an intense beam of neutrinos) and the instruments available could not distinguish the decay of an individual proton from other events that might mimic it. Installing the apparatus 30 meters underground, however, shielded it from most cosmic rays, the chief source of extraneous particles. By assuming all events that could not be accounted for otherwise were due to proton decay, a lower limit of 10^{22} years was set on the proton's lifetime.

Over the next two decades further experiments pushed the lower limit back several more orders of magnitude. Some of the experiments were based on a radiochemical analysis of geologic samples. For example, the disintegration of a proton within a nucleus of potassium 39 converts the nucleus into argon 38, which is unstable and promptly emits a neutron to become argon 37. This last nucleus is radioactive, so that its concentration can be measured by simple radiation counters. The key to the experiment is to find specimens of rock that have remained buried and undisturbed since they formed. From the ratio of argon 37 to potassium 39 one can then deduce the number of protons that have decayed over a geologic period. By this method it was shown that the half-life of the proton is at least 10^{26} years.

Other experiments, more like the ones now under way, attempted to count proton decays as they took place. The measurements were made in conjunction with studies of cosmic rays done deep underground, where all but the most penetrating radiation was screened out. In 1974 one of us (Reines) reported the results of work with a detector set up two miles below the surface in a South African gold mine. The device included 20 tons of liquid scintillator as well as some 84,000 flash tubes, in which an electrical discharge in a gas is triggered by the passage of a charged particle. Again a limit on the lifetime was calculated by isolating those events whose origin was unknown and assuming they could all be attributed to proton decay. The limit set in this way was greater than 10^{30} years, but it applied only to certain decay modes, namely those in which the decay products included either a muon or a positively charged pion. (The muon is a particle related to the electron but with a mass about 200 times larger; the pion is the lightest member of the family of particles called mesons.)

The entire character of the study of proton decay changed with the advent of minimal SU(5). Earlier a few theorists had been enthusiasts for experiments testing proton stability, notably Jogesh C. Pati of the University of Maryland and Abdus Salam of the International Centre for Theoretical Physics in Trieste. Now many more took an interest. In part the change in outlook came from the simple fact of having a definite prediction that the proton does decay. Equally important, however, minimal SU(5) offered an estimate of the lifetime, and one that did not seem beyond the range of possible measurement. The theory also supplied a preferred mode of decay: in the commonest events a proton would break down into a positron and a neutral pion. These predictions gave a goal and a direction to experiments.

The decay mode favored by minimal SU(5) happens to be one that leaves a distinctive signature. The positron and the neutral pion both have comparatively high energy—about .5 GeV—and they fly off in opposite directions. Such "back to back" events at high energy are unlikely to arise from any cause other than proton decay. Hence the events yielding a positron and a neutral pion should be easier to identify than those resulting from many other decay modes.

Even when one knows what to look for, however, testing the predictions of minimal SU(5) is a formidable task. A half-life of 10^{30} years corresponds to the decay of about one proton or neutron per day in 1,000 metric tons of matter. To allow for uncertainties in the theory, an experiment ought to be able to detect the decay even if the half-life is as long as about 10^{33} years. Thus in the worst case 10^{33} protons, or roughly 3,000 tons of matter, must be kept under observation for a year merely to detect a single decay. Furthermore, that decay must be recognized against a background of many extraneous events.

Two strategies can be adopted for dealing with background events: the events can be excluded from the detector or identified and discounted in the data recorded. In practice both techniques are important.

One source of background is radiation from natural radioactivity, which cannot be screened out entirely. Any shielding material chosen would itself include some nuclei susceptible to radioactive decay; for that matter, so would the material of the detector. On the other hand, emissions from radioactive nuclei are relatively easy to identify. The energy of a typical radioactive decay is less than 1 percent of the energy released by the decay of a proton, and so a simple energy measurement serves to discriminate between the two kinds of event.

Cosmic rays are more troublesome. They come in all energies, and at the earth's surface they include a wide variety of particle types. At sea level the flux is about one particle per square centimeter per minute, so that a 10,000-ton detector would be pelted by more than 10^{12} particles per year. To find a single proton decay in the course of a year, a trillion cosmic-ray events would have to be identified and rejected.

It is to reduce the flux of cosmic rays that detectors are put underground. Many of the incident particles, such as protons, neutrons and pions, can be absorbed by a few meters of heavy shielding, but excluding muons calls for more extreme measures. Muons lose energy very slowly as they move through matter, so that for the shielding to be effective it must be thousands of meters thick. Even then a few high-energy muons leak through.

In the case of the neutrino, shielding is quite impossible. Neutrinos in the energy range characteristic of proton decay interact with matter so seldom that they readily pass through the entire earth. Of course the rarity of neutrino interactions implies that most of the neutrinos also pass through the volume of the detector without causing any disturbance, but now and then a neutrino does collide with a particle of matter. The frequency of the events is proportional to the mass of the detector (just as the frequency of proton decays is); in 1,000 tons of material there is one neutrino interaction every few days. In some cases the debris from the interaction includes muons, electrons and pions with energies comparable to what would be expected from proton decay.

Since neutrino-induced events cannot be prevented, they must be distinguished from proton decays. The property that allows the distinction to be made is the angular distribution of the products of the event. When a neutrino strikes a particle, the neutrino's momentum causes the collision products to be thrown forward. In contrast, when a proton at rest decays, the particles emitted move in opposite directions, with a net momentum of zero. The need to make such distinctions has had an important influence on the design of detectors for proton-decay experiments. It is not enough for the detector to record the total energy of the event, as some earlier instruments did; the position and direction of the particles must also be determined.

The detectors planned and built over the past decade are of two general types: layered tracking detectors and water Cerenkov detectors. In a tracking detector plates of iron or steel are interleaved with electronic "counters" sensitive to the passage of a charged particle. The iron provides the stock of protons and neutrons whose decay is awaited. The counters record the successive positions of the decay products as they pass from plate to plate. The spatial resolution of a tracking detector (and hence the accuracy with which a trajectory can be reconstructed) depends in part on the thickness of the plates. Thin plates give better resolution, but they also increase the cost of the device, since more particle counters are needed.

The water Cerenkov detectors are based on an effect discovered in 1934 by the Russian physicist Pavel A. Cerenkov. The effect is the emission of light by a charged particle moving through a transparent medium, such as water, faster than the speed of light in that medium. (Nothing can move faster than the speed light has in a vacuum, but in water light itself is slowed to about three-fourths of its vacuum speed.) The Cerenkov light is an electro-

magnetic shock wave analogous to the sonic boom formed when an aircraft exceeds the speed of sound. Like the sonic boom, the Cerenkov radiation is emitted in a cone. The angle between the particle's direction of motion and the direction of light emission depends on the ratio of the particle's speed to the speed of light in the medium. For a particle moving through water at nearly the vacuum speed of light the angle is 42 degrees.

In a Cerenkov detector a large volume of clear water is surrounded by photomultiplier tubes. Each charged particle created by the decay of a proton or a neutron in the water gives rise to a brief flash of Cerenkov radiation, which causes some of the photomultipliers to "fire," or produce an electrical pulse. The amplitude of the pulses and their times of arrival are recorded on magnetic tape for later analysis. The pattern of photomultipliers exposed to the light and the sequence in which they fire provide the information needed to reconstruct the path of the particle. The photomultiplier tubes must be extraordinarily sensitive. At a distance of five meters the pulse of Cerenkov light from a single charged particle is about as bright as that from an ordinary flashbulb seen at the distance of the moon.

Because Cerenkov radiation is emitted only by charged particles, it might seem that electrically neutral particles would escape unseen. Some of them do—notably neutrinos—but others can be detected through their secondary decay products. A neutral pion, for example, generally decays quickly into a pair of gamma rays, or high-energy photons. The gamma rays are also electrically neutral, but they interact to produce pairs of electrons and positrons, which can be detected. The neutral pion is therefore observed as a cascade of electron-positron pairs.

One advantage of the water Cerenkov detector is that water, being made up partly of hydrogen atoms, includes protons that are not bound up with other particles in a complex atomic nucleus. The signal of proton decay should be clearest in the case of such free protons. Another advantage of the water Cerenkov detector is that its cost does not increase in proportion to its mass. The major cost is not the active medium (water) but the photomultipliers. Since they are mounted at the surface of the enclosed volume, their number depends on the surface area, which is proportional to the two-thirds power of the mass. In short, the cost per unit of mass diminishes with increasing detector size. For this reason the water Cerenkov technique has been adopted for the largest proton-decay detectors. Ulti-

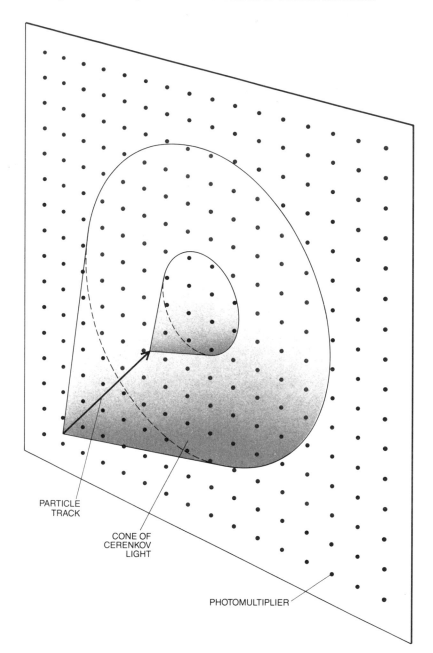

PARTICLE
TRACK

CONE OF
CERENKOV
LIGHT

PHOTOMULTIPLIER

Figure 58 CERENKOV RADIA-TION is emitted when a charged particle moves through water faster than the speed of light in water. Cerenkov light forms a cone centered on the particle's path, and so the path can be reconstructed from a recording of the light's intensity and time of arrival at the photomultipliers in the detector array.

mately the size of such instruments is limited by the absorption of light in water, but the limit has not yet been reached.

We have taken part in the design and operation of the largest of the water Cerenkov detectors. It is a rectangular vat of water 23 meters long,

17 meters wide and 18 meters deep and has a total volume of more than seven million liters. On all six sides photomultiplier tubes are mounted one meter apart on a square grid. There are 2,048 tubes in all (see Figure 59).

Planning for the experiment was begun in 1979 by physicists (including us) from the University of

California at Irvine and the University of Michigan. We were later joined by Maurice Goldhaber of Brookhaven, and so the project came to be known as the Irvine-Michigan-Brookhaven, or IMB, collaboration. The aim of the experiment was to observe proton decay or, failing that, to extend the lower limit on the lifetime to 10^{33} years. The choice of a detector technology was the first major decision. Layered tracking detectors had been in use (for slightly different purposes) for 10 years or more, whereas the concept of a large water Cerenkov detector, capable of recording enough information to trace the path of a charged particle, was as yet untried. Simulations done with a computer, however, showed that the idea was sound.

There remained the difficult question of where to put the detector. At the earth's surface 100,000 cosmic-ray events per second would swamp the electronic system. The flux of cosmic rays would have to be reduced to no more than a few per second, which meant going underground about 2,000 feet. How could such a huge, water-filled apparatus, the size of a six-story building, be assembled at that depth at acceptable cost?

We considered excavating a cavern in a mine or tunnel and building a tank within it, but the cost was too high. We searched for existing caverns, but we found none that were suitable. The solution came from an unexpected quarter. Members of the group had worked on cosmic-ray experiments in a mine operated by the Morton Salt Company (now Morton Thiokol, Inc.) near Cleveland. Through this connection we learned that the management of the mine was interested in testing a new continuous-

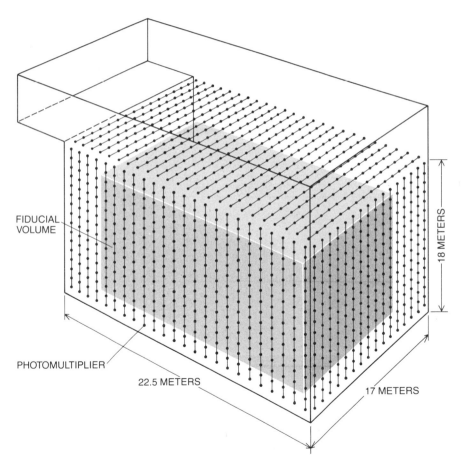

FIDUCIAL VOLUME

PHOTOMULTIPLIER

22.5 METERS

17 METERS

18 METERS

Figure 59 8,000-TON DETECTOR in the Morton Thiokol salt mine. Because of the cosmic-ray background, an event is counted among the potential proton-decay candidates only if its apparent point of origin is at least two meters from the detector's walls. The inner or "fiducial" volume thus defined (color) has a mass of 3,300 metric tons.

mining machine and would share the cost of bringing it underground. More important, they agreed to provide at cost the mining skills and labor needed to excavate the large cavern our experiment required. To reduce the cost further we decided to forgo a freestanding tank and instead line the walls of the cavern with plastic.

The pool was filled by July, 1982, and the installation of the 2,048 photomultiplier tubes was completed soon after. The tubes are arranged along the walls and floor of the cavern and near the surface of the water, facing inward. Each tube has a five-inch hemispherical face and is mounted in a watertight plastic enclosure. The total mass of the water in the cavern is 8,000 metric tons, but events originating in a two-meter shielding zone nearest the walls on all six sides are eliminated from consideration. Excluding the shielding zone leaves a "fiducial" mass, in which candidate proton decays are identified, of 3,300 tons.

An event is recorded whenever 12 or more photomultipliers fire within a period of 50 nanoseconds. This is the time needed for light to travel 10 meters in water, the largest distance likely to be covered by the products of proton decay. The time resolution of the photomultipliers is five nanoseconds, and so the sequence in which the tubes fire can be used to determine the direction from which the light came.

Even at a depth of 2,000 feet the detector is triggered by penetrating muons about 2.7 times per second. In passing through the detector from top to bottom a muon gives off enough Cerenkov radiation to fire about 600 photomultipliers; a proton decay, on the other hand, would fire no more than about 250 tubes. Many of the muon-induced events can therefore be identified and rejected on this basis alone. About a third of the muons, however, cut across a corner of the detector and fire fewer than 300 tubes. These events cannot be rejected until the

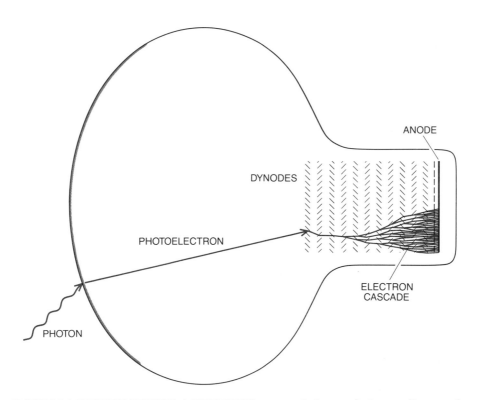

Figure 60 WHEN A PHOTON STRIKES A PHOTOMULTIPLIER TUBE the photocathode on the inner surface of the glass envelope emits an electron, which is attracted to a series of dynodes in a "venetian blind" arrangement, with each row at a progressively higher voltage. A dynode struck by an electron emits several more electrons; as a result the number of electrons is multiplied by a factor of 10^9 and a single photon can give rise to a measurable pulse of current.

recorded data are analyzed. The discrimination is done by a computer program that excludes from further consideration all events initiated by a particle entering the detector from outside. The events remaining, known as contained events, are those originating within the fiducial volume. If examples of proton decay are to be found, they must be among the contained events.

By no means are all the contained events plausible candidates for proton decay. Neutrinos (and occasionally other neutral particles) can enter the detector unseen, then interact with matter to generate an event that appears to originate within the fiducial volume. These events must be rejected by a detailed analysis in which the geometry of the event is the most useful criterion for judgment. As noted above, the debris created when a neutrino collides with a particle of matter usually moves in the general direction of the incident neutrino. In proton decay the commonest expected pattern consists of particles moving in opposite directions.

Unfortunately the distinction between these patterns is not perfectly sharp. On rare occasions one of the products of a neutrino interaction can bounce backward, at a large angle to the direction of the incident neutrino. Moreover, the back-to-back alignment of the products of proton decay can also be disrupted. If the decaying particle happens to be one of the isolated protons that constitutes a hydrogen nucleus, the angle between the products in a two-body decay should be precisely 180 degrees. In the water molecule, however, 16 of the 18 protons and neutrons present are bound in the oxygen nucleus, where scattering of the decay products can weaken their back-to-back angular correlation.

Only a small subset of the contained events satisfy the geometric criterion. These few events are examined further in an analysis drawing on all the information supplied by the detector, including the total energy of the event, the number of particles, the paths they follow and in some cases the kinds of particles emitted. Each event is considered separately as a possible candidate for each hypothetical decay mode of the proton and the neutron. The number of candidates for each mode that cannot be rejected determines the lower limit on the lifetime for that mode. In calculating the lower limits we also take into account the detector efficiency: the percentage of events in each mode that we could expect to recognize in the data, given a large sample.

The members of the IMB collaboration have completed an analysis of the data recorded during 204 days of operating the detector. In that time there were 169 contained events. Are any of them consistent with proton decay? In the case of the distinctive mode favored by minimal $SU(5)$, yielding a positron and a neutral pion, the answer is unequivocal: no events were seen with these particles in the characteristic back-to-back arrangement. The lower limit established by the experiment for decay in this mode is a half-life of 1.7×10^{32} years. The theory had predicted that the lifetime would be found to lie in the range between 10^{28} and 2.5×10^{31} years, and so there is clearly a disagreement. When only this mode is considered, the proton lives at least seven times longer than the predicted maximum.

For other decay modes the results are not as clear-cut. There is no compelling evidence that we have observed proton decay in any mode, but there are a number of events whose classification is to some degree uncertain. They could have been induced by neutrino interactions, but the possibility cannot be excluded that they are genuine proton or neutron decays. In setting limits on the lifetime we have adopted the conservative policy of considering each event a candidate decay unless it can definitely be explained otherwise. The resulting limits are given in Figure 61.

The question of whether the proton decays has not been settled by this experiment. In one respect the situation is much as it was before the introduction of minimal $SU(5)$: experimentalists have no quantitative goal to set the scope of their efforts. The proton lifetime could lie just beyond the range of the present detectors, or it could be many orders of magnitude greater. In another way the situation is quite different. Even though minimal $SU(5)$ appears to be inadequate, it is now generally acknowledged that theoretical considerations imply the instability of the proton. Moreover, observation of proton decay would be considered a unique and persuasive test of the idea of grand unification.

If the decay is to be seen, or if the limits are to be extended, larger and more sensitive detectors will be needed. The existing instruments are being modified and new ones are being built. We are upgrading our own detector by installing larger photomultiplier tubes, and other schemes for increasing the

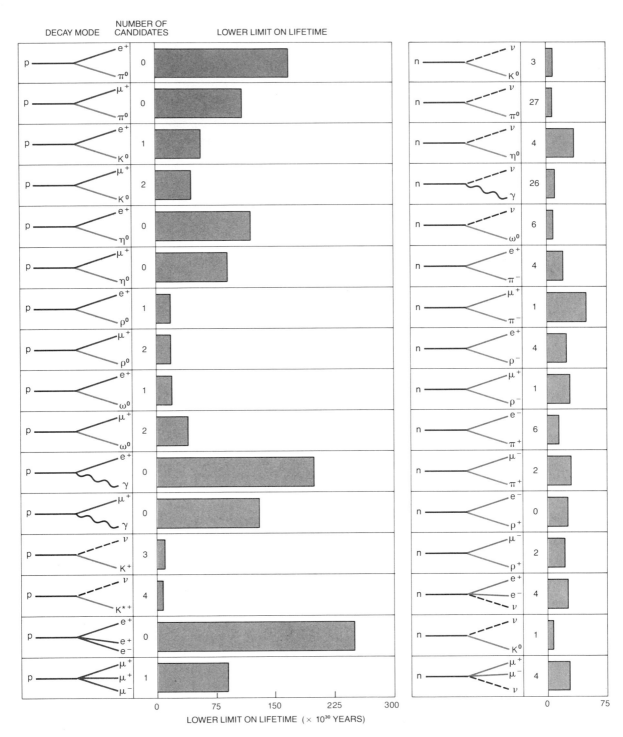

Figure 61 LIMITS ON THE LIFETIME of the proton and the neutron are given for 32 possible decay modes. All events that could not definitely be identified as cosmic-ray interactions were considered possible decay-mode candidates. The minimum lifetime for each mode was calculated from the number of possible candidate events and from the detection efficiency, the percentage of events of each type that the detector could be expected to identify.

amount of light collected are under consideration. The changes will improve our ability to distinguish the various decay modes from the patterns generated by cosmic-ray neutrinos.

Does the proton decay? Will the question ever be answered? The present lower limits on the lifetime can doubtless be extended somewhat, but this process cannot continue indefinitely. Perhaps someday a detector 10 times as large as the IMB device might be built, but a detector 100 or 1,000 times as large is not feasible. Cost is not the only constraint (although it is a formidable one). If the proton lifetime is much greater than 10^{33} years, the irreducible background of neutrino interactions would probably obscure many decay modes no matter how large the detector was. Adding to the mass would simply increase the number of background events in the same proportion as the decay events.

Although the question of proton decay remains unsettled, the history of the search provides an interesting commentary on the relations between theory and experiment. When the question was first raised, the limits on the proton lifetime were not very stringent, and yet most physicists expected the answer to be no. The limits have now been increased by 15 orders of magnitude, but most physicists have come to believe the answer is yes.

POSTSCRIPT

The search for proton decay continues. Our detector now has four times the sensitivity we described. No events uniquely attributable to proton decay have been observed so that the limits on the stability of the proton have improved even further. In the meantime many other detectors have become active and their null results confirm our limits.

Beyond our efforts to detect proton decay, we have now searched for superheavy magnetic monopoles, studied some sources of ultrahigh-energy cosmic rays, investigated many properties of neutrinos and joined in the search for dark-matter candidate particles and supersymmetric particles. This is possible because many of the unique detector properties needed to study proton decay — large mass and heavy shielding — are useful in studying these objects.

One of the most exciting moments in the IMB program came on February 23, 1987, when the detector observed a burst of neutrinos coming from a supernova explosion in the Large Magellanic Cloud galaxy, 160,000 light-years away. The first observation of neutrinos from beyond our solar system and its confirmation by other terrestrial experiments have indeed given us a new window on particle physics in the cosmos.

Superheavy Magnetic Monopoles

Isolated north and south magnetic poles are predicted to exist but have never been observed. A new theory may explain why: such particles seem to be too massive, slow-moving and rare.

. . .

Richard A. Carrigan, Jr., and W. Peter Trower

In 1269 Petrus Peregrinus, an early French investigator of the magnetic properties of materials, described the orientation of bits of iron near the surface of various lodestones. He noted that the lines of force around such a natural magnet are invariably concentrated at two points, just as the meridians of the earth come together at opposite geographic poles. The analogy led him to designate the two points the north and south poles of the magnet. Subsequent observations have confirmed that all ordinary magnetic objects have paired regions of opposite polarity, that is, all magnets are dipoles.

It is easy to conceive of an isolated north or south magnetic pole. Speculation about the possible existence of such magnetic monopoles has persisted for centuries, but there has been no evidence of them. Interest in the idea became more focused in 1931, when the British physicist P. A. M. Dirac showed that an important observed property of electrically charged particles could be explained by assuming the existence of elementary particles bearing a magnetic charge. Dirac's conjecture stimulated a flurry of theoretical papers on the expected properties of the hypothetical monopoles, and several experiments were undertaken to detect them. None of the attempts was successful.

Recent efforts to forge a unified theory of the fundamental forces of nature have again drawn attention to the absence of magnetic monopoles. Indeed, one proposed theory requires the creation of monopoles in the first instants of the great explosion in which the universe was presumably born. The theory goes on to offer an explanation of how the monopoles could exist without having been detected in any of the earlier searches: they would be extraordinarily massive and so would have properties quite different from those of ordinary particles. Several novel experiments are now being prepared to look for superheavy magnetic monopoles among the surviving particles of the big bang.

The dipolar nature of ordinary magnetic materials can readily be demonstrated. If iron filings are sprinkled on a piece of paper held over a bar magnet, the filings trace out a pattern of smooth arcs extending from one end of the magnet to the other. The arcs represent the magnetic field lines between the poles. Where the lines are close together the magnetic field is strong; where they are far apart the field is weak. Cutting the magnet in half does not isolate the poles; instead two smaller bar magnets are created. (It is possible to simulate the field surrounding an isolated magnetic pole by placing the magnet on end, with one pole directly under the

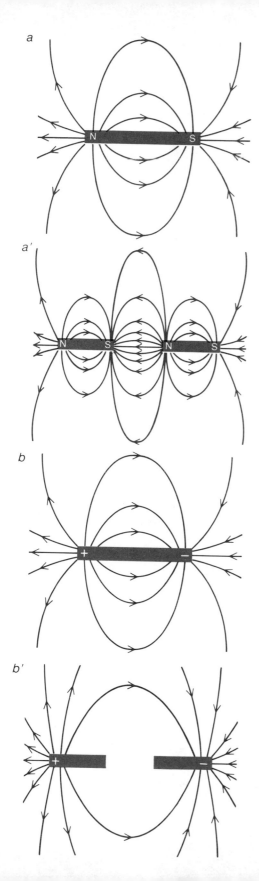

paper; in this case, however, the lines of force connecting the poles are merely removed from the plane of the paper.)

An analogous electric dipole can be made by depositing electric charges of opposite sign on the ends of an insulating rod. When the electric dipole is cut, two isolated electric poles are created. The reason is that each electric pole represents a clustering of individual electrically charged particles: negatively charged electrons at one end and positive ions at the other. When the poles are separated, the aggregations of charge are not affected. The possibility of isolating electric poles but not magnetic ones is a fundamental distinction between electricity and magnetism.

The explanation of this distinction has been known for more than a century. The magnetism of an ordinary object, such as a bar magnet, arises not from clusters of magnetically charged particles but from loops of electric current. For example, the magnetic field of a solenoid (a cylindrical electromagnet) is set up by the electric current circulating in the coil. On a submicroscopic scale similar currents are generated by the circulation of electrons around atomic nuclei. In nonmagnetic materials the atoms, hence the currents, are randomly oriented. If the atoms are somehow aligned, the material shows a net magnetization. In a permanent magnet the atoms retain their alignment even when the orienting force is removed. Cutting such a magnet in half cannot isolate the poles because each atom is effectively a dipole. (See Figures 63 and 64.)

This basic distinction between electricity and magnetism is at the heart of the theory of electromagnetic phenomena formulated by James Clerk Maxwell in 1864. In Maxwell's theory the possibility of isolated magnetic charges was ignored, since none had ever been observed. Instead all magnetism was explained in terms of moving electric charges. Over the past century Maxwell's theory has been put to many experimental tests and has never been found wanting. That fact alone severely limits the contexts in which magnetic monopoles might be found.

Figure 62 DIPOLE FIELDS are set up by a bar magnet (*a*) and by an insulating rod with opposite electric charges deposited at the ends (*b*). When the magnet is cut in half, two smaller dipoles are created (*a'*). When the rod is cut in half, the field remains dipolar because the electric charges that generate the field remain in place (*b'*). Magnetic field lines are in color, electric field lines in black.

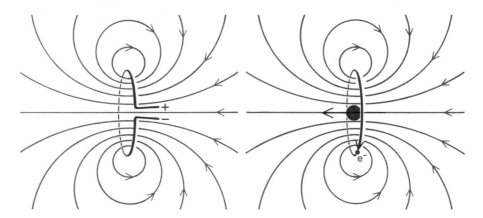

Figure 63 CIRCULATING ELECTRIC CHARGE in a loop of wire generates a dipole magnetic field with its axis oriented at a right angle to the plane of the loop (*left*). The movement of bound electrons around the nucleus of an atom constitutes a similar loop of current and endows the atom with a corresponding dipole field (*right*).

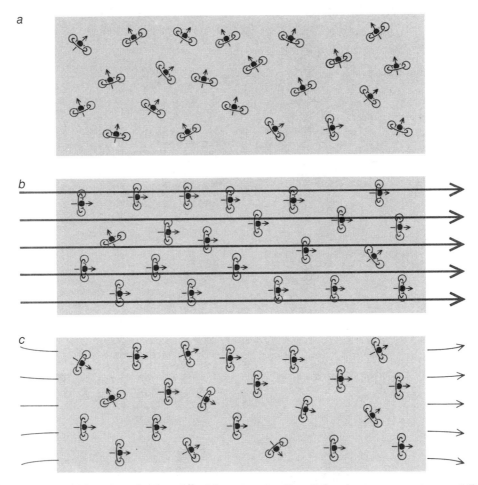

Figure 64 ATOMIC MAGNETS are randomly oriented in an ordinary, nonmagnetic iron bar (*a*). The atoms can be aligned by the application of an external magnetic field (*b*). When the external field is removed (*c*), many of the atoms remain aligned, forming a permanent magnet. For simplicity the bar is represented as having a single magnetic domain.

Dirac's contribution to the subject stemmed from his effort to explain the quantization of electric charge: the observation that electric charge appears only in multiples of the charge of the electron and the proton. Dirac showed that if an isolated magnetic pole exists anywhere in the universe, electric charge must be quantized everywhere. Until recently Dirac's magnetic-monopole hypothesis was the only explanation of the observed quantization of electric charge. (The existence of the particles called quarks, with charges of one-third and two-thirds the magnitude of the electron's charge, would not alter Dirac's conclusion. Many physicists think the quarks must be permanently confined to the interior of particles that invariably have an integer value of net charge.)

The reasoning behind Dirac's proposal can be presented in nonmathematical terms as follows. An atom that has been excited to an elevated energy state tends to revert abruptly to a lower energy state, simultaneously emitting its excess energy in the form of a photon, or quantum of electromagnetic radiation. The photon not only removes energy from the atom but also carries away some of the atom's intrinsic angular momentum. Hence the electromagnetic field represented by a beam of light, say, can be considered to have a certain amount of angular momentum.

It is a fairly straightforward task to calculate the amount of angular momentum in an electromagnetic field. Indeed, more than 80 years ago J. J. Thomson, in a textbook on electricity and magnetism, suggested this exercise for students: Determine the angular momentum of an electromagnetically bound system consisting of a single electric charge and a single magnetic charge. The solution reveals that the angular momentum of the system depends on the product of the electric charge and the magnetic charge but is independent of the distance between them. In other words, the electric charge and the magnetic charge may be separated by the radius of an atom or by the radius of the universe; in both cases the angular momentum of the electromagnetic field is the same.

Now, the angular momentum of any system of this kind is known to be quantized: the smallest amount of angular momentum in nature is equivalent to the unit called Planck's constant, and all larger amounts are multiples of this unit. If it is assumed that the angular momentum of the system is equal to some number of Planck units and that the magnetic charge has a definite value, then it

follows that the electric charge is also fixed. In this way Dirac was able to show mathematically that if magnetic charge exists, electric charge must be quantized.

Dirac's theory incorporated a curious mathematical construct he called a string. A Dirac string resembles an infinitely long solenoid with a magnetic monopole at one end; the rest of the string threads off into the distance. The string is an impediment to detailed calculations based on Dirac's model. Recent developments in mathematical physics, however, have made it possible to divest the monopole of its troublesome tail.

Dirac's quantization condition endows the magnetic monopole with certain properties. For example, to account for the observed quantization of the angular momentum of an electromagnetic field the minimum unit of magnetic charge must be about 70 times as large as the corresponding unit of electric charge. According to another prediction of Dirac's, every particle of matter, including the magnetic monopole, must have an antimatter counterpart. Just as the electron has its antiparticle (the positron, discovered in 1932, four years after Dirac foretold its existence), so the magnetic monopole is expected to be matched by a magnetic antimonopole. In keeping with Peregrinus' arbitrary nomenclature, one particle of such a pair is called a north monopole and the other a south monopole. Dirac's theory made no prediction about the mass or size of the magnetic monopoles or about their abundance in the universe.

Some interesting effects arise when Maxwell's equations of electromagnetism are augmented to include magnetic charges and magnetic currents. For example, as the velocity of a moving electric charge approaches the speed of light, its properties should increasingly resemble those of a magnetic charge; similarly, a moving magnetic monopole would begin to take on the properties of an electric charge at a speed approaching the speed of light. These transformations, which follow from Einstein's special theory of relativity, have been confirmed experimentally for moving electric charges but not of course for moving magnetic charges.

A moving electric charge can lose energy by ionizing matter (that is, by detaching electrons from their atoms). Typically, energy is lost at the rate of a few million electron volts for every centimeter traveled through a substance. The energy required to ionize an atom is generally a few tens of electron volts, and so a moving electric charge can ionize hundreds of thousands of atoms per centimeter.

Because of the much stronger charge of the magnetic monopole, it would ionize atoms some 10,000 times more effectively. Thus a magnetic monopole passing through a photographic emulsion of the type employed by physicists to detect electrically charged particles would leave a track thousands of times darker than the track left by an electric charge moving at the same speed. Because the monopole would lose energy to the ionization process so quickly, it would slow down much sooner on entering a substance than an electrically charged particle with the same kinetic energy does.

Just as an electric field can accelerate an electrically charged particle, so a magnetic field could accelerate a magnetic monopole. Because of the greater pole strength of the magnetic particle, however, it would gain energy faster in a magnetic field than an electrically charged particle does in an equivalent electric field. A magnetic monopole traversing a superconducting coil one meter long would gain more energy than a proton acquires in the largest particle accelerator yet built.

The physics of magnetic monopoles has another curious feature, which can be made apparent by imagining that the flow of time could be reversed. In a thought experiment suggested by Robert K. Adair of Yale University a proton is moving through a magnetic field, which causes it to follow a curved path. In one case the magnetic field is produced by an electric current in a coil. The effect of reversing time is to reverse both the motion of the proton and that of the electrons making up the current; hence the magnetic field is also reversed. Under these circumstances the proton simply retraces the same path in the opposite direction; the path of the proton is said to be invariant with respect to time reversal.

Now suppose the magnetic field arises not from an electric current but from the presence of a magnetic monopole. Reversing time does not alter the polarity of the monopole and therefore leaves the direction of the magnetic field unchanged. As before, the proton reverses direction, but it does not retrace its path. In short, the proton's path in the field of a monopole depends on the direction of time, an effect that violates the principle of time-reversal invariance.

The predicted effects of a magnetic monopole when time is reversed were for many years viewed as a serious argument against its existence. In 1964, however, Val L. Fitch, James W. Cronin and their colleagues from Princeton University, in an experiment done at the Brookhaven National Laboratory, discovered an effect much like a violation of time-reversal invariance in the decay of the particles called neutral kaons. The full theoretical significance of this finding has only recently begun to be understood. As this understanding has increased, some of the opposition to the idea of magnetic monopoles has abated.

Given the tantalizing properties of magnetic monopoles, what is the status of the experimental search for evidence of their existence? Soon after every new particle accelerator is commissioned magnetic monopoles are looked for in the debris of the initial high-energy particle collisions; searches of this kind have become virtually a rite of passage. Monopoles have also been sought among the by-products of collisions between cosmic rays and atoms in the atmosphere. Experiments of another kind have attempted to detect monopoles among the iron atoms of terrestrial and extraterrestrial substances. So far none of these searches has succeeded.

How are the experiments done? One approach is to look for monopoles in iron that has been exposed to a beam of high-energy particles from an accelerator. If monopoles are produced by such a beam, some of them should bind to the iron by inducing an opposite magnetic charge in the material (much as a note-holding magnet, say, clings to the door of a refrigerator). A powerful electromagnet should then be able to pull the monopoles out of the iron. Monopoles liberated in this way would be detected in particle counters designed so that only very strongly ionizing particles would be recorded. Samples of iron ore collected from the rocky outcroppings of old mountains are another potential source of material for this method of extracting monopoles. Indeed, this approach was initially taken in the 1940's by Willem Malkus of the University of Chicago.

Another detection method, first discussed in the 1960's, was implemented in the 1970's by Luis W. Alvarez and his colleagues at the Lawrence Berkeley Laboratory of the University of California. In their device a sample of material suspected of harboring magnetic monopoles is passed repeatedly through a superconducting coil. On each pass of a magnetic monopole the electric current in the coil would presumably increase by a small amount. Because the

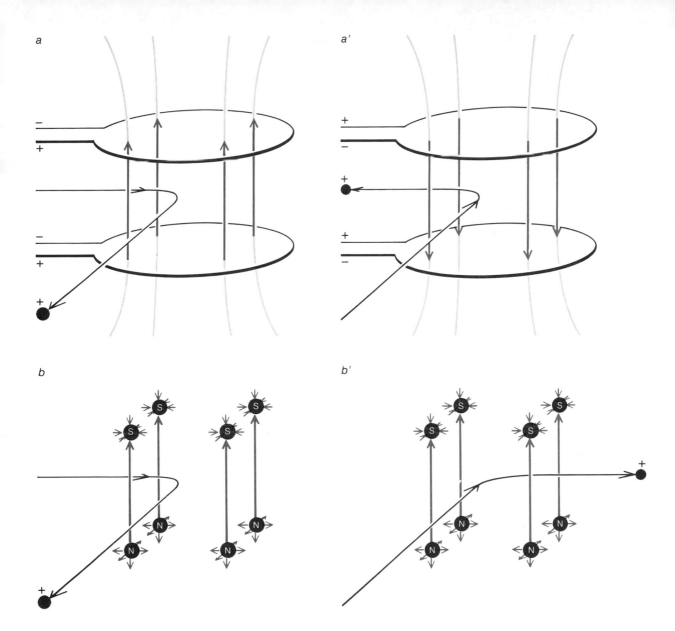

Figure 65 TIME REVERSAL. A proton moves along a curved path through a perpendicular magnetic field generated by electric currents flowing in a pair of wire loops (*a*). If the direction of time is reversed (*a'*), the currents and the motion of the proton would be reversed. In an idealized array of north and south monopoles (*b*), reversing time would leave the magnetic field unchanged (*b'*). Although the proton would reverse direction, it would not retrace its path.

coil is superconducting the incremental induced current would persist indefinitely. The task then becomes one of measuring the extremely small signal induced by multiple passes of a single monopole. By means of this technique Alvarez and his colleagues were able to show that the density of magnetic monopoles in rock samples recovered from the surface of the moon is less than one for every 10^{28} protons. Even at this limiting abundance, however, there could still be an average of one monopole in every 20 kilograms or so of matter.

A less direct way of hunting for magnetic monopoles is to look for signs of the creation and destruction of a monopole-antimonopole pair. In theory a

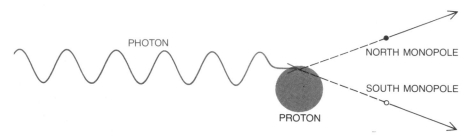

Figure 66 PARTICLE-ANTIPARTICLE PAIR, consisting of north and south monopoles, could be created when a high-energy photon, or quantum of electromagnetic radiation, interacts with an electrically charged particle such as a proton. The mutual attraction between the monopoles, however, would cause them to collide, converting their mass back into photons.

pair of this type could be created when a high-energy photon passes near a proton, just as an electron-positron pair is known to be produced (see Figure 66). The oppositely charged monopoles would exist for only a moment, however. They would immediately be attracted to each other, bending their paths in a way that would cause them to spew out photons of *bremsstrahlung*, or "braking radiation." They would soon come together and annihilate each other, converting their mass into additional photons.

This hypothetical mechanism was invoked by Malvin A. Ruderman and Daniel Zwanziger of New York University in the mid-1960's to explain a few unusual cosmic-ray events recorded in the late 1950's. Each of the events consisted of a single jet of tightly bunched, very energetic, photon-induced showers of electrons in which there was no evidence of any strongly interacting nuclear particles. Two subsequent experiments done with the large proton accelerator at the Fermi National Accelerator Laboratory (Fermilab) found no evidence of showers of this type. (One of the experiments was done by a group from the Virginia Polytechnic Institute and State University and the other by a group from the University of Michigan.) If the process is real, it might take place only at energies higher than those currently available in particle accelerators. Alternatively, the process could be much rarer than the cosmic-ray evidence suggests.

In 1975 the world of physics was jolted by the announcement that a magnetic monopole had been discovered. The claim was made by investigators at the University of California at Berkeley and the University of Houston. Their evidence was an anomalously thick, dark track, presumably of cosmic-ray origin, recorded in a stack of photographic emulsions and plastic sheets. The detector had been exposed to cosmic rays while it was suspended from a balloon flown at high altitude for two and a half days. Soon after the announcement the interpretation of the event as evidence of the passage of a magnetic monopole came under widespread criticism. A cosmic-ray experiment of this kind is characterized by its area-time factor: a measure of the area of the detector multiplied by the exposure time. The detector in which the candidate track was seen had an area-time factor roughly a million times smaller than that attained in previous searches in which no monopole had been seen. In the analysis of moon rocks by Alvarez and his colleagues, for example, the rocks had been exposed to particles of all kinds for billions of years.

Other problems with the monopole interpretation of the event subsequently led the experimenters to suggest instead that the track might have been caused by the passage of a superheavy atomic nucleus or a massive antiparticle. One benefit of the episode is that it has inspired a careful evaluation of the pivotal question of how a magnetic monopole would lose energy through ionization. Even so, the question remains unsettled.

At about this time the prospects for magnetic-monopole hunters suddenly brightened. Working independently, Gerard 't Hooft of the University of Utrecht in the Netherlands and Alexander M. Polyakov of the Landau Institute for Theoretical Physics near Moscow found that a certain class of theories of elementary-particle interactions not only allow magnetic monopoles but also demand them. Furthermore, the new theories, which are called gauge theories, indicate that the monopoles must be much more massive than any particle previously seen or even predicted. In retrospect it was hardly surprising that none had been detected up to then.

Besides having an extraordinarily large mass, the magnetic monopoles proposed by 't Hooft and Polyakov differ in several other respects from the original Dirac monopoles. For one thing, the 't Hooft-Polyakov monopoles do not require a string. For another, they are not pointlike particles, although they are expected to be too small to be directly measured.

The superheavy 't Hooft-Polyakov monopoles play an important role in attempts to construct a "grand unified theory" to describe three of the four known forces of nature. Two of these forces, the electromagnetic force and the weak nuclear force, are already linked by the highly successful electroweak theory, which treats them as different manifestations of a single underlying force. The goal of most of the current attempts to create a grand unified theory is to arrive at a more comprehensive mathematical structure that would incorporate both the electroweak force and the strong nuclear force (omitting only gravity, the fourth known force).

One of the most interesting consequences of such grand unified theories is the prediction that the proton will decay into other particles. To account for such a disintegration novel particles called leptoquarks are introduced to change quarks (the supposed constituents of particles such as the proton) into leptons (particles such as the electron that respond to the weak nuclear force but not to the strong one). The leptoquarks would be extremely heavy, perhaps 10^{14} times as heavy as the proton. If, as the 't Hooft-Polyakov hypothesis suggests, a superheavy magnetic monopole is associated with the leptoquarks, the monopole would have a mass of about 10^{16} proton masses, or roughly 20 nanograms. This is an exceedingly large mass for an elementary particle, comparable to that of a paramecium or an amoeba.

Superheavy monopoles could have been created only in the first 10^{-35} second after the birth of the universe. The big bang of creation would have been the only event hot enough (almost 10^{30} degrees Kelvin) to generate such particles. Both north and south magnetic monopoles would have been formed, and a small fraction of them would have recombined, annihilating each other. Most of the superheavy monopoles would have escaped an early death, however, and there is no reason to think they would not have survived to the present.

It is unclear where the monopoles would have collected as the universe evolved, but then it is also unclear how the universe evolved from the big bang into the galactic structures we see today.

One feature of our own galaxy makes it possible to set a stringent limit on the number of magnetic monopoles that could be present locally. The galaxy has a magnetic field, and although it is weak (on the order of a hundred-thousandth of the earth's magnetic field), it extends over enormous distances. A monopole caught in the galactic field would be accelerated to very high energy and would eventually escape from the galaxy's gravitational attraction. Because the monopoles would be extremely massive, however, they would still move slowly (at velocities of a few thousandths the speed of light).

Eugene N. Parker of the University of Chicago has pointed out that if there were too many monopoles of this kind in the galaxy, they would destroy the galaxy's magnetic field. His argument is based on the fact that the magnetic field of the galaxy is generated by the large-scale circulation of electrically charged particles. As a magnetic monopole is accelerated it would drain energy from the galactic field by slowing the currents of moving electric charges. The existence of a galactic magnetic field therefore places an upper limit on the population of magnetic monopoles in the galaxy.

The maximum density of monopoles in the universe is related to the unresolved question of whether the universe will continue to expand forever or will eventually collapse. The outcome depends on the amount of matter in the universe. The amount of visible mass (that is, the mass of luminous objects such as stars) is not quite enough to cause the expansion to slow, stop and ultimately reverse. Unless there is some additional, unseen mass the expansion of the universe will gradually slow, but it will never stop.

Is there enough invisible matter in the universe to affect its fate? What form of matter could be present in great abundance and yet remain undetected? One possibility is that the missing mass consists of neutrinos, which rarely interact with other particles and never give off light. The neutrino was long thought to be massless, but there is now much speculation that it does have a small mass. Because neutrinos are thought to fill the universe with an average density of approximately a million per cubic centimeter, even a small neutrino mass could make a significant contribution, although probably not much more than the visible mass.

In all likelihood superheavy magnetic monopoles would radiate little light and hence would also be an invisible component of the universe. If the monopoles have a mass 10^{16} times the proton's mass, it would not take many of them to greatly augment the total mass. At a concentration of just one monopole for every 10^{16} protons there would be roughly as much mass in monopoles as there is in luminous matter. It does not seem feasible for the total mass of the universe to be much more than 10 times the visible mass, and so as an upper limit the ratio of magnetic monopoles to protons must be less than one to 10^{15}.

In 1979 John P. Preskill, then a graduate student at Harvard University, combined the grand unified model of the strong, weak and electromagnetic forces with standard cosmology to argue that there should be about one magnetic monopole for each proton in the universe. From the analysis of the expansion rate of the universe, on the other hand, it appears that there should be less than one monopole per 10^{15} protons. Preskill formulated a dilemma: Either the role of magnetic monopoles in the grand unified theories is wrong or standard cosmology is wrong. One way out of Preskill's dilemma is to adjust the cosmological model, allowing more monopole-antimonopole annihilations in the early moments of the universe. Another option is to suppress the estimated initial production rate of magnetic monopoles by some unspecified mechanism.

More recently George Lazarides, Qaisar Shafi and Thomas Walsh of the European Organization for Nuclear Research (CERN) in Geneva considered how the predicted density of superheavy magnetic monopoles in the universe could be reduced by adjusting either particle theory or cosmological theory. They concluded that the interaction of monopoles with the galactic magnetic field sets a limit on the ratio of magnetic monopoles to protons of about one to 10^{20}. Given that abundance, some 200 monopoles per year would be expected to pass through an area of one square kilometer. A more conservative estimate, based on a more uniform distribution of monopoles in the universe, would result in a flux of a few monopoles per year per square kilometer.

For the first time the theory of magnetic monopoles thus provides estimates of the expected mass and flux of magnetic monopoles. Armed with these estimates, however rough they are, the experimenter now has a fresh field to explore. The predicted flux of magnetic monopoles is small, but not so small that it is out of the question to look for them.

One place to look for superheavy monopoles is in large-scale natural effects. Indeed, this is the direction originally suggested by Parker a decade ago. One of us (Carrigan) has speculated on the fate of monopoles in the material that accreted to form the solar system. As the earth condensed, for example, magnetic monopoles would have sunk toward the center under the influence of the planet's gravitational and magnetic fields. North monopoles would have collected near the south geomagnetic pole and vice versa.

From the geologic record it is known that the earth's magnetic filed has reversed many times. Such a field reversal would cause the two separated populations of monopoles to migrate toward and then through each other. During their journey some monopoles and antimonopoles would be annihilated, liberating the enormous energy embodied in their mass. From the measured heat flow at the surface of the earth one can set a rough limit on the number of monopoles trapped in the core; the number calculated in this way is consistent with other experimental limits on the abundance of superheavy monopoles.

A more straightforward approach is to construct a detector specifically to search for these heavy, rare particles. The design of such a detector, however, is not obvious. Indeed, the art of searching for massive monopoles is now at one of those engaging moments in science when a wealth of ideas, many of them quite bizarre, are at war on paper and over lunch tables. Massive monopoles are expected to travel slowly, at speeds far below the velocity of light. Just what would happen when a slow-moving monopole struck an atom is not clear.

The collision of a superheavy monopole and a stationary atomic nucleus would be like a steamroller hitting an ant. A cosmic-ray monopole could lose a huge amount of energy to many such encounters as it plowed its way ponderously through the earth, and it might still emerge virtually unscathed from the other side. Under these circumstances it is difficult to predict what degree of ionization would be observed in a detector. One view holds that there are enough fast-moving monopoles for detectors relying on ionization to be used in the search. The other view is that ionization will be seen rarely and only weakly, so that unconventional techniques would be needed to detect monopoles. In either case, an extremely large detector is clearly needed if the experimenter is to observe a monopole event in his lifetime.

One detector that records the light generated by ionization and covers many square kilometers has been developed by Haven E. Bergeson, George L. Cassiday and Eugene C. Loh of the University of Utah. The device, called the fly's-eye detector, is an array of photomultiplier tubes directed at the night sky; it registers the light given off by secondary particles produced by rare ultrahigh-energy cosmic-ray interactions in the upper atmosphere. As the secondary particles shower down toward the earth they collide with nitrogen atoms in the atmosphere, causing them to scintillate. The fly's-eye detector can distinguish this light only if the total energy released in the event is equivalent to at least 100 million proton masses. The passage of a magnetic monopole, even with the most optimistic estimate of its ionization rate, would give rise to less than a ten-thousandth of the light needed to set off the detector.

The ability of such a detector to respond to particle-induced scintillations is limited by background illumination from stars, overflying aircraft and other sources such as the beacon lights of distant radio towers. One suggestion for alleviating the problem of background light is to mount a detector on the rim of the Grand Canyon, point it into the canyon and record data only on cloudy nights. Even this measure, however, would reduce the background light only tenfold. In addition it might be difficult to get hikers to forgo campfire meals for several years. Perhaps a fly's-eye detector could be installed in a large cave or a salt mine such as those now being used to look for proton decay.

Another large-volume detector now being planned is the Deep Underwater Muon and Neutrino Detector (DUMAND), which will be sensitive to events within a cube of ocean about a kilometer on a side [see "A Deep-Sea Neutrino Telescope," by John G. Learned and David Eichler; SCIENTIFIC AMERICAN, February, 1981]. DUMAND will respond to the Cerenkov radiation emitted when a particle moves through the seawater faster than the speed of light in water. Unfortunately superheavy magnetic monopoles would probably move too slowly to give off Cerenkov radiation.

Some of the largest existing scintillation detectors, such as the giant neutrino detectors at Fermilab and CERN, are too small by a factor of about 100 to have a good chance of observing magnetic monopoles if the flux is limited by the galactic magnetic field. Nevertheless, it may still be useful to turn these detectors to searching for monopoles during accelerator shutdowns, when otherwise they would be unemployed, since the experimental limits for slowmoving, superheavy particles are not well defined. A preliminary search by Jack D. Ullman of Lehman College in New York employed a detector half a meter square operating for several months, thereby providing the only experimental limit obtained to date. Definitive experiments will have to be 10,000 times as sensitive.

The contrary view holds that all searches with ionization detectors are doomed to failure because the slow-moving, superheavy monopoles will cause no ionization. The passage of any charged particle through metal is accompanied by eddy currents, however, regardless of the particle's speed and of whether its charge is electric or magnetic. The eddy currents can be expected to create acoustic pulses, which might be detected. Carl W. Akerlof of the University of Michigan has checked the possibility of constructing a spherical metal detector along these lines and has concluded that a signal could be detected above the background of thermal noise only if the detector is cooled to a few millidegrees above absolute zero. This difficult technical requirement is compounded by the necessity of making the detector large enough to sense the low expected flux of monopoles.

One comparatively simple strategy for detecting superheavy monopoles calls for a superconducting coil similar to the one employed by Alvarez and his colleagues. The detector is called a SQUID (for superconducting quantum interference device), and it registers changes in an electric current when a free cosmic-ray monopole passes through it. Blas Cabrera of Stanford University is currently searching for monopoles with a superconducting niobium coil five centimeters in diameter. Cabrera also has a second monopole search under way. In the latter experiment he inflates a cylindrical, superconducting lead bag one meter long and 20 centimeters in diameter, expelling most of the trapped magnetic flux. If a magnetic monopole penetrates the bag, it will leave trapped magnetic flux in the parts of the wall where it entered and left. By mapping the magnetism of the bag periodically, differences in the flux patterns can be attributed to the passage of a monopole. If a monopole is detected, rough information on its direction will be recorded.

David B. Cline of Fermilab and the University of Wisconsin at Madison and Carlo Rubbia of CERN and Harvard University have considered an ambitious monopole detection effort. They plan to mount a superconducting detector under an iron-ore processing plant in Wisconsin. The plant heats more than a million tons of ore per year to a temperature of 1,700 degrees Celsius. At this temperature any magnetic monopoles trapped in the iron would be released, allowing them to fall through the detector.

The story of the magnetic-monopole conjecture is unlike any other in physics. Begun half a century ago by one of the giants of modern physics, the hunt for monopoles has been a fertile field for theoretical speculation but to date it has been barren of any supporting experimental evidence. The discovery of a magnetic monopole would rank as one of the finds of the century, comparable to the discovery of the positron, Dirac's other great prediction. If the monopole were found to be very massive, the case for some form of grand unified theory of elementary-particle interactions would be strengthened.

In the more likely event that no magnetic monopoles are found, the negative evidence will continue to be viewed as inconclusive. The necessary experiments will be difficult, and even if they are perfectly executed, a null result will not bring much clarity to the situation. Nevertheless, the ultimate vindication of the monopole idea rests in these searches, since physics is in the final analysis an experimental science.

POSTSCRIPT

Nearly simultaneously with the publication of our article Cabrera was observing a candidate monopole signal in his detector. Meanwhile Alan H. Guth's theory of the inflationary universe, which we only hinted at, took aim at reducing the number of monopoles produced in the early universe. Inflating the cosmos, discussed in Chapter 11, "The Inflationary Universe," by Guth and Paul J. Steinhard, causes only a few monopoles to remain in the universe, a discouraging possibility for a hunter. Inflation explains Preskill's dilemma as well as the apparent homogeneous character of the universe. Further, sophisticated theoretical calculations soon showed that a monopole passing through matter could catalyze a proton to decay.

Cabrera builds extremely good magnetic shields, creating spaces essentially free of magnetic fields. He had attached a 5-centimeter-diameter loop to a SQUID inside this shield to measure precisely another important quantity, h/m, the ratio of the Planck constant to the mass of the electron. During six months when the detector wasn't in active use, it sat waiting for a magnetic monopole to thread its loop. On St. Valentine's Day in 1982 the device recorded a current change whose value was exactly that expected from a monopole. Cabrera's event was very clean in a detector uniquely suited to a monopole search. A second but less convincing candidate event was seen at Imperial College in the summer of 1985 in another induction detector.

In some sense, Cabrera had no business looking for monopoles as ionization detectors had already set lower limits than Cabrera could reach for free cosmic-ray monopoles based on their expected high ionization. However, ionization experiments were open to criticism; slow-moving massive monopoles would not be so strongly ionizing, and so might escape detection.

A greater reason for suspicion was the Parker bound. Even a small taint of free monopoles in the galaxy would have shorted out the galactic magnetic field, which is measured to be a few microgauss. The allowed monopole flux was 10,000—100,000 times lower than suggested by the Cabrera event.

In short order a three-front attack was mounted on the problem of monopole abundance. First, the underlying principles of the ionization detectors were reexamined, and it was shown that some electronic arrangements should be able to detect slowly moving monopoles. Old and new detectors have swung into action, quickly establishing plausible limits that in total now approach the Parker bound. Japanese physicists have also been processing tons of iron ore while a Berkeley group has set interesting limits with fossil fission fragment tracks. Second, the astrophysicists have had a field day—assessing the Parker bound, looking at the effect of monopole catalysis on neutron stars and generally setting astonishingly low limits on monopole abundance. Finally, a new wave of larger inductive detectors, typically 10 to 100 times larger than Cabrera's original device, have been developed. These multiple-loop coincidence detectors require less shielding but have more inductive coupling between the shield and the loop. Newer inductive experiments have reduced the limits of the monopole flux an order of magnitude per year from that of the original Cabrera event. This is still 100 times higher than the Parker astrophysical bound.

What is the future for magnetic monopole searches? The original impetus to develop inductive detectors was stimulated by the Cabrera candidate and a suspicion that slow monopoles might not register in ionization detectors. Larger inductive detectors are very expensive while analyses of the ionization properties of monopoles show that efficient electronic detectors can be made. Indeed, many of the electronic monopole hunters have combined forces to fill one gallery in the Italian Gran Sasso Underground Laboratory with a football field–sized detector that should alone exceed the Parker bound annually — a worthwhile goal.

But doesn't inflation and the lack of convincing experimental confirmation rule out magnetic monopoles? No. There is as yet no completely satisfactory link between inflationary cosmology and particle physics. This is further complicated by the lack of experimental verification of the most straightforward grand unified theory. Newer physics — supersymmetry or superstrings — still can have built-in monopoles. Real fun for the future would be if one of these theories predicted a laboratory-sized monopole.

PARTICLES AND THE UNIVERSE

...

Deuterium in the Universe

All the heavy hydrogen in space may have been made in the first 15 minutes after the big bang. Observations leading to estimates of its abundance thus provide evidence on conditions at that time.

. . .

Jay M. Pasachoff and William A. Fowler

The calculations of modern cosmology lead to the conclusion that the universe originated between 10 and 20 billion years ago. The particular cosmology that seems to have the greatest weight of observational evidence behind it is the "big bang" theory, which posits that all the matter in the universe was compressed into a superdense kernel that exploded and has been expanding ever since. Within the past few years evolving techniques of observation have made it possible to look backward in space and time to what appear to be the first few minutes after the big bang, and to learn about the physical state and early evolution of the universe.

One item of observational evidence in favor of the big-bang theory was the discovery that the universe is permeated uniformly by radiation corresponding to what would be radiated by a theoretically perfect "black body" with a temperature of three degrees Kelvin (degrees Celsius above absolute zero). This radiation is believed to be a remnant of the original big bang. The newest evidence has come to the fore only recently. It is the detection in interstellar space of atoms that may have been formed within the first 15 minutes after the big bang. These atoms are the atoms of deuterium, commonly known as heavy hydrogen.

The big-bang theory stemmed from the discovery by Edwin P. Hubble half a century ago that all other galaxies appear to be receding from ours, and that the most distant galaxies seem to be receding the fastest. The big-bang theory accounts for Hubble's observations by the expansion of the universe, as does the steady-state cosmology advanced by Hermann Bondi and Thomas Gold and independently by Fred Hoyle. The steady-state theory, however, postulates that the universe has always looked the same as it does today, and it requires that hydrogen be continuously created to make up for the decreasing density of the universe caused by its expansion. Most astronomers believe this explanation has been ruled out by the discovery of the three-degree background radiation. A variant big-bang cosmology posits that the universe will stop expanding and then contract. The recent measurements of interstellar deuterium have provided a method for determining whether that will happen or whether the universe will simply go on expanding forever.

The "standard" big-bang theory states that in the first 100 seconds after the initial explosion of the universe only protons, neutrons, electrons, positrons and various types of neutrinos existed and that each particle was independent. As the exploding universe cooled, the protons and neutrons began to combine. Since the nucleus of an atom of "ordinary" hydrogen is simply a proton, the combinations of protons and neutrons were the nuclei of the heavier isotopes of hydrogen and of the other atoms. The simplest compound nucleus is the deuteron, composed of one neutron and one proton, which with a single electron forms an atom of deuterium. The nucleus of the third isotope of hydrogen, tritium, consists of one proton and two neutrons. The combination of two protons and one neutron forms the nucleus of an atom having a quota of two electrons; it is the light isotope of helium. The nucleus of the common isotope of helium has two protons and two neutrons. The nuclei of heavier elements are built up in the same way.

In 1948 Ralph A. Alpher, Hans Bethe and George Gamow first calculated the way in which the nuclei of the lighter elements could have been synthesized in the big bang. In 1957 one of us (Fowler), together with Geoffrey Burbidge, E. Margaret Burbidge and Hoyle, laid out a blueprint showing how both the lighter and the heavier elements could have been built up in the interior of stars. Seven years later Hoyle and Robert J. Tayler calculated that the large amount of helium observed throughout the universe (between 20 and 30 percent of the total mass) could not have been manufactured in ordinary stars. The nuclei require a temperature of 10 billion degrees K. for their synthesis, a temperature that could have been attained only in the big bang, in the explosions of supermassive stars ("little bangs") or in supernovas, the explosions of ordinary stars.

In 1966, after the discovery of the three-degree background radiation, P. J. E. Peebles tried to calculate what the relative abundance of deuterium and helium would have been if those two elements had been formed in the big bang. At the same time Robert V. Wagoner joined Fowler and Hoyle to check whether or not the new data sufficiently changed the outcome of previous calculations enough to allow heavy elements to have been synthesized in the big bang. They found that it was still not possible. Could these elements have been formed in events other than the big bang? Let us deal first with their synthesis in the big bang and then discuss the possible exceptions.

The calculations of Wagoner, Fowler and Hoyle show that there is only one isotope left over that is unique to the big bang. It is deuterium. Although helium would also have been made in the big bang, additional helium has been synthesized since then

in stars and cannot now be distinguished from the original helium. Deuterium, however, is only depleted by the processes that go on inside ordinary stars; it is "cooked" into heavier elements by thermonuclear reactions. Therefore whatever deuterium we find now may well have persisted since the origin of the universe. The deuterium would have been formed in the first 1,000 seconds after the big bang, when the rapidly expanding universe had cooled off just enough to allow protons and neutrons to combine. In some sense, then, deuterium may actually allow us to look backward 10 to 20 billion years in time.

In order to use deuterium as a key to gaining a fuller understanding of the big bang, it is necessary to determine how abundant deuterium is with respect to ordinary hydrogen. Ideally we should like the cosmic abundance ratio, if it exists as a meaningful constant, to be uncomplicated by processes inside stars or by the chemical interactions of elements and compounds on planets or on grains of ice or dust in space. The ratio is best measured in the gas between the stars.

The density of the interstellar gas is very low: about one atom per cubic centimeter. The space between the stars is so vast, however, that the total amount of gas is very large. It is well known that almost all this gas is hydrogen. The problem is to find out if any of the hydrogen is deuterium and if so how much.

How can deuterium be observed? In an atom of ordinary hydrogen the single proton and single electron act as though they are spinning like tiny tops. The spin of the electron can be either in the same direction as the spin of the proton or in the opposite direction. If the spins are in the same direction, the atom is in a higher energy state, and the electron spin can flip over so that it is spinning in the opposite direction with respect to the proton. This spin-flip emits a small amount of energy at the radio wavelength of 21 centimeters. Gaseous hydrogen can also absorb energy at this wavelength as well as emit it. The atoms that absorb the energy are thus put into the higher energy state. Although the spin-flip for any one atom is rare, there are enough atoms in the galaxy for the radiation at 21 centimeters to be quite strong.

Deuterium also has a single electron, and the radiation from its spin-flip is at a wavelength of 92 centimeters. Deuterium, however, is much less abundant than ordinary hydrogen. In the water of the earth's oceans there are 6,600 atoms of ordinary hydrogen for every atom of deuterium. The abundance of deuterium in the solid body of the earth could be quite different. Thus the value for the oceans does not necessarily reflect the abundance ratio of deuterium to ordinary hydrogen on the earth, much less the ratio in interstellar space.

Soon after the spectral line of hydrogen at 21 centimeters was discovered, several astronomers began to search for the line of deuterium in interstellar space. The surveys culminated 10 years ago when Sander Weinreb of the National Radio Astronomy Observatory (NRAO) at Green Bank, W.Va., turned the observatory's 85-foot radio telescope to the constellation Cassiopeia to look for the 92-centimeter line in the spectrum of the strong radio source Cassiopeia A. Although Weinreb ob-

served for weeks, he was totally unable to detect deuterium. He stated on the basis of his negative observations that the ratio of deuterium to ordinary hydrogen in that direction appeared to be less than 1:13,000, since any larger amount would have been detected. That upper limit is half the ratio of deuterium to ordinary hydrogen on the earth.

Radio astronomy has continued its rapid development since the time of Weinreb's survey, and within the past few years many complex molecules have been detected in the clouds of gas and dust in interstellar space. Most of the molecules have been found at relatively short wavelengths, from about 21 centimeters down to two millimeters. Few spectral lines have been discovered at wavelengths longer than a meter, although the failure has not

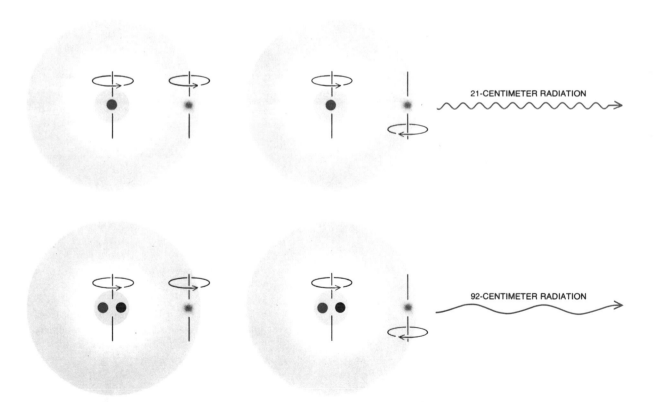

Figure 67 SPIN OF THE ELECTRON flips on rare occasions in an atom of hydrogen or deuterium with the emission of radiation. In the schematic at top left the nucleus and electron in an atom of ordinary hydrogen spin in the same direction. When the electron flips so that it is spinning in the opposite direction (*top right*), radiation is emit-ted at a wavelength of 21 centimeters. The same kind of spin-flip in the deuterium atom (*bottom drawings*) gives rise to 92-centimeter radiation. When the atoms absorb radiation of same wavelength, the spin of their electron flips back.

been for lack of trying. In 1969 Carl A. Gottlieb, Dale F. Dickinson and one of us (Pasachoff) used the 150-foot radio telescope of the Air Force Cambridge Research Laboratories to search at meter wavelengths for various molecules, some of which included deuterium as a constituent. Two such molecules were the hydroxyl radical (OH), with deuterium in place of ordinary hydrogen, and water, with deuterium in place of one of the two ordinary hydrogens. The search was later continued at the NRAO in collaboration with David Buhl, Patrick Palmer, Lewis E. Snyder and Ben Zuckerman. No molecules were detected.

The chance of finding the faint spectral line of deuterium, or for that matter any other spectral line at such wavelengths, seemed very small. Nevertheless, in conversations in 1970 the authors of this article felt that another attempt should be made. The project would call for many weeks of observing, but it seemed that the 130-foot radio telescope of the Owens Valley Radio Observatory of the California Institute of Technology might be available. Diego A. Cesarsky and Alan T. Moffet of the staff at Owens Valley joined one of us (Pasachoff) in making the observations. In the following discussion of these observations "we" refers to Cesarsky, Moffet and Pasachoff and students from Williams College and Cal Tech who worked with us.

We decided to observe for two weeks at a time because we wanted to analyze the results as we went along. Many problems could have interfered with the investigation. For example, the 92-centimeter line lies in a region of military air-to-ground communications. Although the aircraft transmissions were sometimes bothersome, we were able to remove those time periods from our data.

The first two weeks of observing were in March, 1972. The main direction in which we chose to observe was toward the center of our galaxy. The galactic center, however, is in the southern sky, and at the latitude of the Owens Valley it is above the horizon for only about six hours a day. Since radio telescopes can operate in broad daylight, we were able to observe the galactic center whenever it was above the horizon, to observe in the direction of the Great Nebula in Orion much of the rest of the time and to observe in the direction of Cassiopeia A during any intervening hours. The data from this first observing run looked good, showing that the telescope and the receiving system were working satisfactorily. Furthermore, there was actually a suspicion of a faint absorption line at 92 centimeters.

That summer we had three more observing runs of two weeks each, extending the results of the first run. We concentrated our efforts in the direction of the galactic center. There is more hydrogen in front of that source than in any other part of the sky. We assumed as a working hypothesis that the abundance ratio of deuterium to ordinary hydrogen is constant throughout the universe, so that presumably the greatest total amount of deuterium would lie in the same direction as the greatest total amount of ordinary hydrogen. The galactic center is a strong source of radio waves, and the gas in the 40,000 light-years between it and the earth absorbs some of the radiation at the wavelengths of 21 to 92 centimeters.

Our analysis of the data by computer during that summer and fall strengthened our belief that we were actually observing the absorption line of deuterium. In the summer of 1973 five more weeks of observing by Cesarsky and Pasachoff yielded results that were compatible with the findings in 1972. We attempted to observe over a wider range of wavelengths, but it seemed as though we were now encountering too much outside interference in the wider band. In the narrower range of wavelengths observed in common during both years, the largest absorption fluctuation again corresponded to the spectral line of deuterium at 92 centimeters.

The line is still only barely visible in the data, which show expected random fluctuations (see Figure 68). Adding the results of both years' observations together has slightly improved the ratio of the signal to the noise and the line seems more than three times as strong as the fluctuations. Even if the "line" turns out to be only a particularly large random fluctuation in the data, the magnitude of the surrounding noise places an important upper limit on how strong the absorption could be. From our data that upper limit is one part in 3,000. If the feature we observed is the deuterium line, then the abundance ratio of deuterium to ordinary hydrogen would be between 1:3,000 and 1:50,000.

Curiously the major part of the uncertainty in that range arises not because the absorption line of deuterium is so weak but because the radiation from ordinary hydrogen with which it is compared is so strong. In the distance between the galactic center and the earth there is so much hydrogen that the radio signal is saturated. The result is that one cannot tell exactly how much hydrogen there is in this space, and one can only determine a lower limit.

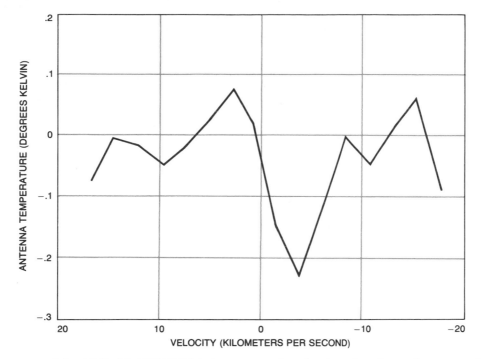

Figure 68 ABSORPTION LINE OF DEUTERIUM shows up as the central dip in observations of the galactic center at a wavelength of 92 centimeters. The dip is shifted in frequency by a small amount, indicating that the gas incorporating the deuterium is moving at a velocity of 3.7 kilometers per second toward the solar system. The spectrum is calibrated in terms of temperature of antenna, equivalent to the energy of radiation received by antenna.

The next step in improving the determination of the abundance ratio of deuterium to ordinary hydrogen is not only confirming the detection of the absorption feature at 92 centimeters but also improving the measurements of the strength of the hydrogen radiation at 21 centimeters.

The gas seems to have the small velocity of 3.7 kilometers per second in our direction. The spectral lines of other gases observed in the direction of the galactic center, notably the hydroxyl radical and formaldehyde (H_2CO), are also Doppler-shifted by an amount that corresponds to the same velocity.

At the same time that we were making our first observations Keith B. Jefferts, Arno A. Penzias and Robert W. Wilson of the Bell Laboratories were observing DCN, that is, the deuterated form of the hydrogen cyanide molecule (HCN). With the 36-foot radio telescope of the NRAO located on Kitt Peak in Arizona, Jefferts, Penzias and Wilson turned their attention to the Kleinmann-Low nebula, a small cloud in the Great Nebula in Orion in which many interstellar molecules have been de-

tected. Within the Kleinmann-Low nebula the DCN molecule was radiating very strong spectral lines at the wavelengths of two millimeters and four millimeters. The observations indicated that the ratio of DCN to HCN was 1:170.

This ratio of DCN to HCN is much higher than the ratio of HDO to H_2O in the earth's oceans. It was immediately realized, however, that if the result was to be understood, certain facts of chemistry would have to be taken into account in addition to the facts of physics. Chemical combinations of the elements follow complicated principles and processes, many of which are unknown. That is particularly true for elements combining on the surface of dust grains in interstellar clouds, which is a leading possibility for the way in which such molecules form. The ratio of deuterium to ordinary hydrogen in interstellar space could very well be quite different from the ratio of DCN to HCN. Calculations show that the difference in the ratios from such chemical fractionation may be on the order of a factor of 600. When the observation of Jefferts,

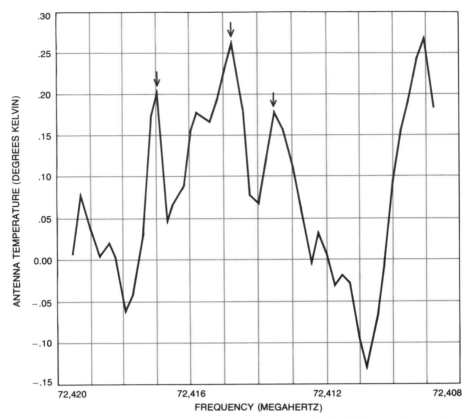

Figure 69 EMISSION LINE OF DEUTERATED HYDRO-GEN CYANIDE (DCN), which is actually composed of three separate lines (*arrows*), was observed at millimeter wavelengths in Kleinmann-Low nebula in constellation Orion with a 36-foot radio telescope on Kitt Peak.

Penzias and Wilson was adjusted to take the correction factor into account, the ratio of deuterium to ordinary hydrogen in the Kleinmann-Low nebula was found to be 1:100,000, just below the range deduced from the observations of Cesarsky, Moffet and Pasachoff. The fact that the ratio of deuterium to hydrogen is indeed substantially lower than the ratio of DCN to HCN has since been confirmed by two independent limits set on the abundance of deuterium by Cesarsky observing at radio wavelengths and by another group observing a transition of deuterium at visible wavelengths.

The abundance ratio of deuterium to ordinary hydrogen has also been determined from space, from the third Orbiting Astronomical Observatory satellite, named *Copernicus*. One experiment, conducted by a group from Princeton University including Lyman Spitzer, Jr., Jerry F. Drake, Edward B. Jenkins, Donald C. Morton, John B. Rogerson, Jr., and Donald G. York, uses a 32-inch telescope to observe spectra at the ultraviolet wavelengths between 950 and 1,450 angstroms and between 1,650 and 3,000 angstroms. The second range includes spectral lines from molecular hydrogen (H_2), detected only once before, in an experiment placed aboard a sounding rocket. Molecular hydrogen had not been detected from the earth's surface because it has no lines in the visible region of the spectrum, and its radiation in the ultraviolet region is absorbed by the atmosphere.

The results from the Princeton experiment aboard *Copernicus* showed that whenever the telescope was pointed toward reddened stars (that is, in directions where the interstellar material in front of stars affects the overall distribution of the energy radiated by the stars by favoring the longer —redder—wavelengths), the spectra revealed that at least 10 percent of the intervening matter was in the form of molecular hydrogen. The same result was obtained with 11 different stars. For a similar number of unreddened stars, however, molecular hydrogen was not detected at all to a limit of one part in 10 million.

Furthermore, the Princeton group measured two lines of HD (deuterated H_2) at wavelengths of 1,054 and 1,066 angstroms. The first results, for nine stars, indicate that the abundance ratio of HD to H_2 is 1:1,000,000. Again the result had to be adjusted by calculations of chemical combinations to yield a corrected ratio of deuterium to ordinary hydrogen. John H. Black and Alexander Dalgarno of the Center for Astrophysics of the Harvard College Observatory and the Smithsonian Astrophysical Observatory have calculated that the ratio of deuterium to ordinary hydrogen is between 1:5,000 and 1:500,000 for the nebula Zeta Ophiuchus.

After these promising measurements had been made Rogerson and York attempted to use the telescope aboard *Copernicus* to observe the Lyman series of transitions of the deuterium atom. These transitions absorb radiation in the ultraviolet region of the spectrum when the electron in the deuterium atom is raised to a higher energy level from the "ground" state. The wavelengths of the transitions are slightly shorter than the wavelengths of the corresponding transitions for the atom of ordinary hydrogen (and were also slightly shorter than the wavelengths for which the experiment was actually designed). The calculation of the ratio of deuterium to ordinary hydrogen from these measurements is straightforward and not subject to the correction factors that must be applied to the molecular observations.

Rogerson and York searched for the transitions in the interstellar gas in front of hot stars of spectral Type B. Such stars have few lines in their own spectra and therefore any absorption lines detected would have been formed in interstellar space between the star and the earth. In the direction of the star Beta Centauri, Rogerson and York were able to measure four lines in the Lyman series for deuterium. The ratio of deuterium to ordinary hydrogen in that direction is 1:70,000, with an error of 15 percent. The work is being continued for other stars. In addition there have been a number of other estimates, as shown in Figure 71. David C. Black of the Ames Research Center of the National Aeronautics

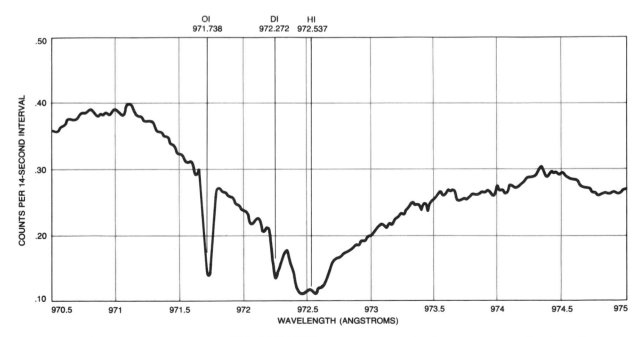

Figure 70 SPECTRUM OF THE STAR BETA CENTAURI shows Lyman-gamma absorption lines for ordinary hydrogen (HI) and deuterium (DI). The lines are from interstellar gas between the star and the solar system. They are superposed on a broad strip Lyman-gamma line from the star (*gentle dip*). A line of oxygen (OI) also happens to be in this region of the spectrum.

and Space Administration deduced the abundance of deuterium in the protosolar nebula from which the sun evolved from measurements of deuterated water (HDO) in meteorites (water in which one deuterium atom replaces one of the two hydrogen atoms in H_2O). Dennis J. Hegyi of the Bartol Research Foundation collaborated with Nathaniel P. Carleton and Wesley A. Traub in their measurements of deuterium in the Great Nebula in Orion at visible wavelengths. Other measurements have been carried out by Mark A. Allen and Richard Crutcher of the California Institute of Technology, Harmon Craig of the University of California at San Diego, G. Boato of the University of Chicago and Nicola Grevesse at the Institute of Astrophysics at Liège in Belgium. "Recombination line" in Figure 71 refers to a spectral line emitted at a radio wavelength when an electron recombining with a deuterium ion to form un-ionized deuterium passes from the 93rd to the 92nd energy level. "Lyman lines" are the spectral lines emitted by a hydrogen or deuterium atom when the electron drops from higher energy states to the ground state. "From ^3He" refers to an abundance of deuterium deduced from the abundance of helium 3.

The big-bang theory assumes that the universe is isotropic (that it looks more or less the same in all directions), so that a knowledge of how uniformly deuterium is distributed through space bears on whether or not the big bang is indeed a good model for the origin of the universe. If the abundance of deuterium is found to be nonuniform, for example by comparing the observations from *Copernicus* of nearby stars with our observations toward the galactic center, then either the big bang is not a good model or the deuterium was formed in some way other than the big bang.

In order for the measurements of deuterium to be significant in cosmological calculations, the measurements must reflect a general cosmic abundance and not simply local variations. For this reason it is difficult to interpret the abundance of deuterium in the solar system in terms of a big-bang origin. That abundance should nonetheless be discussed briefly if only because deuterium had not been detected elsewhere in the solar system until last year.

S ome deuterium has been found in the water content of carbonaceous meteorites. Until the Apollo astronauts landed on the moon, meteorites were the only samples of matter from space. One of the Apollo experiments was designed to capture ionized atoms from the solar wind: the flux of ions expelled from the sun. The astronauts caught the ions in a "window shade" of aluminum foil they unrolled on the moon. The foil was then brought back to earth for analysis. Johannes Geiss of the University of Bern and Hubert Reeves of the Nuclear Research Center at Saclay and the Institute of Astrophysics in Paris made deductions about the abundance of deuterium from the light isotope helium 3 that had been found in the foil. It is known that the sun has transformed into helium 3 most of the deuterium that was present in the primordial solar nebula, the cloud of dust and gas out of which the sun formed. Therefore the amount of helium 3 in the foil directly reflects an upper limit for the amount of deuterium that could have been present in the solar nebula. Geiss and Reeves have calculated that the abundance ratio of deuterium to ordinary hydrogen in the solar nebula was 1 : 40,000.

It is also possible that the deuterium in the material from which the planets were made was not transformed into helium 3 as the protosun materialized out of the solar nebula. In March, 1972, Reinhard Beer of the Jet Propulsion Laboratory of the California Institute of Technology and his collaborators reported that they had detected CH_3D, a deuterated form of methane (CH_4), in the spectrum of Jupiter at infrared wavelengths. Beer and Frederic W. Taylor calculated the molecular-correction factors and found that the abundance ratio of deuterium to ordinary hydrogen in Jupiter was between 1 : 13,000 and 1 : 35,000, depending on certain assumptions about the structure of the planet's atmosphere.

Although the interpretation of results from such a relatively complicated molecule as CH_3D is very uncertain, the calculations seem to show a clear discrepancy with the value of 1 : 6,600 found on the earth. Within the past year John T. Trauger and Frederick L. Roesler of the University of Wisconsin and Nathaniel P. Carleton and Wesley A. Traub of the Center for Astrophysics have detected three lines of deuterated molecular hydrogen (HD) on Jupiter in the near-infrared region of the spectrum. From their observations they calculate a ratio of deuterium to ordinary hydrogen of 1 : 48,000, with a possible error of 20 percent. The discrepancy between the values on the earth and those on Jupiter might be due to the fact that the ratio of heavy water (HDO) to ordinary water (H_2O) on the earth could be enhanced over the ratio of deuterium to

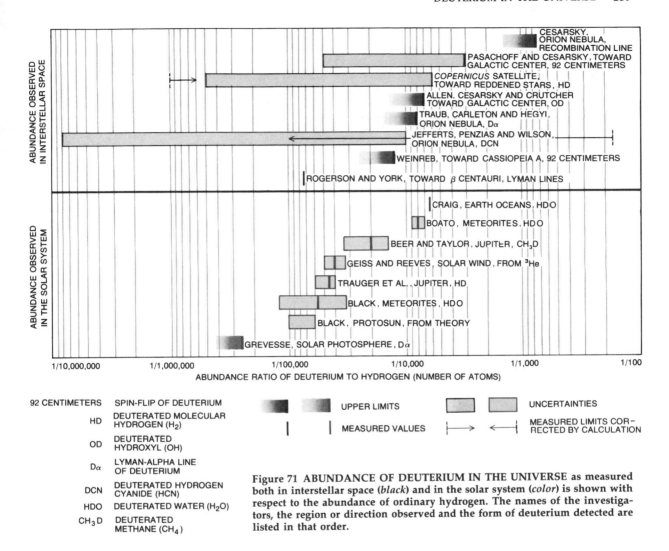

Figure 71 ABUNDANCE OF DEUTERIUM IN THE UNIVERSE as measured both in interstellar space (*black*) and in the solar system (*color*) is shown with respect to the abundance of ordinary hydrogen. The names of the investigators, the region or direction observed and the form of deuterium detected are listed in that order.

ordinary hydrogen by some chemical fractionation process.

A most unusual observation of deuterium was made two years ago. A group of workers from the University of New Hampshire were observing the sun during the large solar flares of August, 1972, with a gamma ray instrument aboard Orbiting Solar Observatory 7. During the particularly large flare of August 7 they detected a peak in the gamma ray spectrum corresponding to the energy that would have been released by the formation of deuterons. That peak meant deuterium was being synthesized on the surface of the sun. This event, however, is probably unrelated to the total cosmic abundance of deuterium. The amount of deuterium formed in this way is small, and it may well be quickly consumed by further nuclear processes on the sun before it can be expelled into space.

The results of the observations of deuterium up to 1974 are summarized in Figure 71. What do these indications of the amount of deuterium in interstellar space tell us about the early universe? One can link the ratio of deuterium to ordinary hydrogen to the big bang by theoretical calculations. For this purpose we shall adopt the standard model of the big bang developed by Wagoner, Fowler and Hoyle, which accepts three basic assumptions about the universe today: first, the universe is homogeneous and isotropic; second, the principle of equivalence must hold, that is, a gravitational field cannot be distinguished from an acceleration (a principle fundamental to the theory of relativity); third, the three-degree background radiation was indeed generated by the big bang. The model also makes several other assumptions. The baryon number must be positive, that is, the amount of antimatter is not equal to the amount of matter, ruling out theories to the contrary. The lepton number is small, that is, the flux of neutrinos does not overwhelm the amount of radiation. The general theory of relativity provides the correct theory of gravitation. Lastly, only the kinds of subatomic particles that we now know about are present; there are no new ones.

In the standard model it happens that the amount of deuterium formed just after the big bang is sensitive to the density of the universe at that time. If the density were relatively high in the first few seconds after the big bang, the deuterium just formed would have been quickly cooked into helium and the end result would be less deuterium. Conversely, if the universe was less dense, the end result would have been more deuterium (see Figure 72). Thus from the measured value of the deuterium in the universe today it is possible to find the density of the universe at the time that the elements were first synthesized. Knowing the density of the universe then and the rate of expansion now, it is possible to calculate what the density of the universe is at present. The density we deduce from our observations is of the order of 10^{-31} gram per cubic centimeter. That density is not enough to "close" the universe, which will thus continue to expand indefinitely. Therefore knowing the universe's present density enables us to draw conclusions about its present state and eventual future.

There are two other ways to measure the density of the universe. The first entails simply adding up the masses of everything that can be observed (stars, galaxies and so forth) and dividing the total mass in a given unit volume by that volume. The most recent work along this line has been done by Stuart L. Shapiro of Princeton. His results depend somewhat on the value accepted for the Hubble constant, that is, on the rate at which the universe is expanding. The value of the Hubble constant is currently a topic of debate in some circles. The traditional value is 75 kilometers per second per megaparsec, which corresponds to a density of 10^{-31}

gram per cubic centimeter. (One megaparsec is 3.3 million light-years.) A newer value measured by Allan R. Sandage of the Hale Observations and Gustav Tammann of the University of Basel is 55 kilometers per second per megaparsec, which yields three-fourths the density of the former value.

The second method of measuring the density of the universe considers the dynamics of interactions of clusters of galaxies; it predicts a density that is substantially higher than the value obtained by the first method. Both values for the density could be correct if there is a substantial amount of invisible material in the universe. This "missing mass" could be in the form of molecular hydrogen in the intergalactic material, although the measurements from the *Copernicus* satellite seem to indicate that there is not enough molecular hydrogen to account for it. Alternatively, the mass could be in the form of the enigmatic black holes. The observed value for the abundance of deuterium in the universe can place an independent limit on how much invisible matter there could be.

The amount of missing mass in the universe touches on a major question in cosmology: Will the universe go on expanding forever, or is there enough matter in it for mutual gravitational attraction to eventually reverse the expansion? In cosmological terms, is the amount of missing mass large enough to "close" the universe? There is a parameter q_0 that is used in cosmological equations. It is inversely proportional to the Hubble constant, and it represents the rate at which the expansion of the universe is slowing down. If q_0 equals 0, the expansion is not slowing down and the universe is open. If q_0 equals 1, then the universe is closed and will eventually collapse on itself. The dividing line between an open universe and a closed universe is a q_0 of $\frac{1}{2}$. In the steady-state model of the universe q_0 is -1.

Sandage has published a "formal" value for q_0 of $.96 \pm .4$, which would mean that the expansion is slowing down and the universe is closed. His value is based on direct observation of the distant galaxies. The uncertainty in the value, however, means that a q_0 of $\frac{1}{2}$ is not ruled out; $.96 - .4$ is $.56$, or

Figure 72 "STANDARD" MODEL of the big-bang theory of the origin of the universe predicts an abundance of deuterium that is very sensitive to the postulated density of the universe at the time it originated, as is shown in this diagram devised by Wagoner. The cosmic abundance of a number of other light elements are also given.

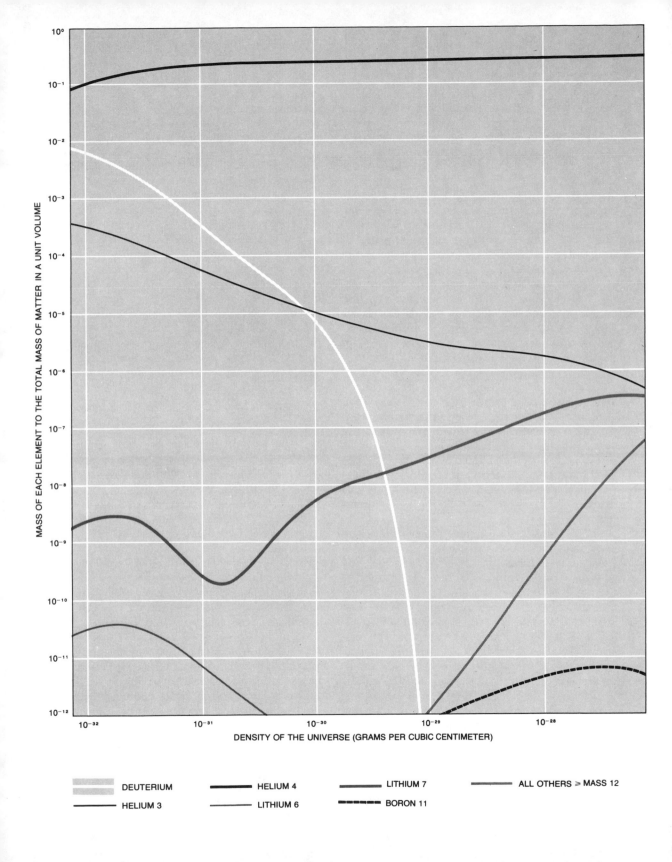

MASS OF EACH ELEMENT TO THE TOTAL MASS OF MATTER IN A UNIT VOLUME

DENSITY OF THE UNIVERSE (GRAMS PER CUBIC CENTIMETER)

DEUTERIUM   HELIUM 4   LITHIUM 7   ALL OTHERS ⩾ MASS 12

HELIUM 3   LITHIUM 6   BORON 11

virtually ½. On that basis the universe would be open. If the deuterium originated in the big bang with an abundance ratio of the levels now being detected, then q_0 is less than .1 and the universe would definitely be open. It is interesting that according to Wagoner's calculations that is the case for the standard model. There are other cosmological methods of assessing q_0 that also yield low values for the density.

There are several ways to get around the possible disagreement between Sandage's value of q_0 from observations of the distant galaxies and Wagoner's value of q_0 from measurements of the abundance of deuterium. Beatrice Tinsley of the University of Texas has hypothesized that because galaxies evolve over a period of time, an observed value of q_0 equal to 1 could mean that the real value of q_0 is 0. The essential point of her argument is that as we look deeper into space we are also looking backward in time, and that the average brightness of galaxies long ago may be quite different from what it is today. Such a fact would distort the scale of distances, which is based on galaxies having a unique average brightness, and thus the value of q_0 would also be distorted. Tinsley's proposal is a controversial one. Sandage has shown that in order for it to account for the discrepancy between his observations and Wagoner's calculations, the galaxies would have to be decreasing in brightness by about .1 magnitude per billion years, which is very fast according to many theories of the evolution of galaxies.

Another way to resolve the disagreement would be to accept the notion that significant amounts of deuterium were formed after the big bang. Recently one of us (Fowler) and Hoyle have calculated that deuterium could have formed in space by a spallation reaction: if a shock wave could inject helium nuclei into an ionized gas composed mainly of hydrogen, deuterium could be knocked out of the helium nuclei by the force of their impact on the nuclei of ordinary hydrogen. Alternatively, if only a neutron were knocked out of the helium nucleus, the neutron could combine with a proton in the surrounding gas to form a deuteron. Deuterium that came about in this way would not readily break up again; since the temperature of the gas would be low, the amount of energy in it would not be large enough to cause the nuclei to dissociate. Exactly how important such a spallation process would be on the galactic scale remains to be seen. Possibly the process might enhance the ratio of deuterium to ordinary hydrogen in certain local regions such as in the Orion nebula and not in others such as the galactic center.

Another mechanism for producing deuterium apart from the big bang has been proposed by Stirling A. Colgate of the New Mexico Institute of Mining and Technology. The observed deuterium could have been synthesized if 3 percent of the mass of the galaxy had at some time been recycled in supernovas. Deuterium could also have been synthesized in supermassive stars, as was suggested by Fowler and Hoyle. Accepting the hypothesis that deuterium was formed apart from the big bang would probably require a number of significant changes in our views of the relative importance of such phenomena as supernovas in galaxies.

Cosmological problems never seem to admit of easy solutions. With the current observational and theoretical work on deuterium moving ahead so rapidly, however, new lines of reasoning and inquiry are opening up for tackling the most basic questions of the universe.

POSTSCRIPT

In 1974, when our article appeared, the time at which deuterium was formed (three minutes or so after the big bang) was considered early in the universe. Our present understanding of the universe has been pushed so that we talk of times 10^{37} times earlier than the time of deuterium formation.

We observed deuterium because, uniquely, no stellar processes produce deuterium when they form other elements. While many theoretical and observational developments have modified the framework of observations, none has changed the idea that the deuterium abundance can reveal the density of baryons in the universe.

We wrote our article after we had reported a possible detection of deuterium from its fundamental spin-flip line, while others had seen it in hard-to-interpret data from molecules and ultraviolet observations in interstellar space. We had made our observations over many months with a Caltech radio telescope. Our attempt to extend these observations with the Parkes radio telescope in Australia was unsuccessful because we needed extraordinary stability, which was not then available with some of their associated equipment. Subsequently, a group in India reached roughly our sensitivity at the spin-flip wavelength. Their spectra showed no deuterium line, but their beam had a different shape from ours and may not have looked at the same region of space.

The molecular observations continued partially mapping the distribution of deuterium-bearing molecules throughout our galaxy. These observations revealed apparent variations in the deuterium abundance, but the correction factors to give the deuterium-to-hydrogen ratio were too large to allow the density of the universe to be reliably calculated.

The largest observational effort has been to study the deuterium component in Lyman spectral lines taken with the *Copernicus* satellite. The first observations seemed to give 1.4×10^{-5} for the deuterium-to-hydrogen ratio and perhaps solve the problem. Indeed, the possibility that the deuterium: hydrogen ratio was then known may well have prevented us from getting the telescope time we needed to continue our radio work. Years of interpretation of *Copernicus* results followed, studying a dozen stars within 1 kiloparsec (about 3,000 light years) of the sun. Eventually apparent variations in the ratio were found from place to place, even within this small region of the galaxy. By 1983 some of the ratios measured with *Copernicus* had fluctuated from hour to hour, impossible for cosmic deuterium. It turned out that only a narrow band around the expected deuterium wavelengths had been measured in some cases and that this was contaminated by Doppler-shifted hydrogen for a few stars. It took several years before it was found that the values for most of the stars were not contaminated.

Deuterium observations have also been made from the International Ultraviolet Explorer spacecraft, though its spectrograph does not have the spectral resolution of the now defunct *Copernicus*. The latest ultraviolet observations, mainly from *Copernicus*, indicate variations of two from place to place in local regions of our galaxy. Thus, we must continue to observe deuterium all the way to the galactic center, some ten times further away.

One of our limitations had been the size of our beam at the 92-centimeter wavelength that results from the deuterium spin-flip. The availability of synthesis telescopes held out hope for working with a narrower effective beam. In early 1987 receivers at the proper frequency had been installed on 16 of the radio telescopes at the Very Large Array (VLA) of the National Radio Astronomy Observatory, located west of Socorro, New Mexico. An array of this type is best suited for finding enhancements of deuterium in small regions, and K. Anantharanaiah, Donald A. Lubowich and I had a test run of one night to search for such enhancements in the direction of the center of our galaxy. Our preliminary result is that no enhancement by a factor of 100 exists. Also in 1987, a negative result was reported for a search for emission from deuterium in the opposite direction from earth, that of the anticenter of our galaxy. Further, astronomers in the Netherlands have begun deuterium observations with an array of radio telescopes there.

Other observations continue to extend deuterium studies. Deuterium molecules have been found in the atmospheres of Jupiter, Saturn, Uranus and Saturn's moon Titan, though the chemical corrections to the pure deuterium-to-hydrogen ratio are too uncertain to allow cosmological interpretation. A current conclusion is that some of this deuterium came from the protosolar nebula.

The theoretical density of the universe evaluated using the deuterium abundance data has not drastically changed since our article. It has, though, incorporated limits on the number of neutrino types in the universe.

The current interstellar deuterium values indicate that the density of baryons is less than $1/10$ of that needed to "close" the universe. How does this square with the prediction of the inflationary universe theory that has the universe on the boundary between being open and closed? If we accept both the inflationary universe and the baryon density from deuterium, we need nonbaryonic matter to provide the additional 90 percent of the mass. Candidates for such nonbaryonic dark matter abound, as discussed in Chapter 1, "Dark Matter in the Universe," by Lawrence M. Krauss. Further observation of deuterium would be useful in evaluating the situation. And we must follow the new theoretical studies of deuterium in an inhomogeneous universe that may explain the observed amount of deuterium at different densities than previously thought.

The Cosmic Asymmetry
Between Matter and Antimatter

It seems the universe today is almost entirely matter. Evidence from both cosmology and particle physics (the study of the universe on the largest scale and the smallest) now suggests an explanation.

. . .

Frank Wilczek

All the fundamental constituents of matter come in matched pairs: for every kind of particle there is an antiparticle that is identical in mass but opposite in other properties, such as electric charge. The symmetrical pairing of particles and antiparticles is required in order to unite the two great theories of 20th-century physics: relativity and quantum mechanics. The symmetry has been well verified by experiment. Since 1932, when the positron, or antielectron, was discovered, the catalogue of antiparticles has grown apace with the catalogue of ordinary particles. Indeed, a particle and its antiparticle have often been discovered simultaneously when the two were created as a pair by a high-energy collision in a particle accelerator. Such collisions always seem to yield matter and antimatter in equal quantities; indeed, it was long assumed that the laws of nature express no preference for matter or antimatter.

And yet in the world outside the laboratory antimatter is almost never encountered. The atoms composing the earth consist of neutrons, protons and electrons, but never their antiparticles. Does this asymmetry prevail throughout the universe?

That is, does the entire universe consist predominantly of matter, with very little antimatter? If it does, has the asymmetry always existed, or did the universe begin with equal numbers of particles and antiparticles and somehow develop an imbalance later?

Recent findings in cosmology and particle physics suggest answers to these questions. They suggest that in the first instant after the big bang, when the universe was much hotter and denser than it is now, there were equal amounts of matter and antimatter. Before the universe was 10^{-35} second old, however, violent collisions among particles created conditions that led promptly to an asymmetry between matter and antimatter. The asymmetry has been locked into the universe ever since. The road leading to this conclusion is still unpaved in places, but I shall try to show that the route is the right one.

How can one be sure the universe consists entirely of matter? It is easy to demonstrate that matter and antimatter cannot be mixed homogeneously. Whenever a particle and the corresponding antiparticle come together, they annihilate each

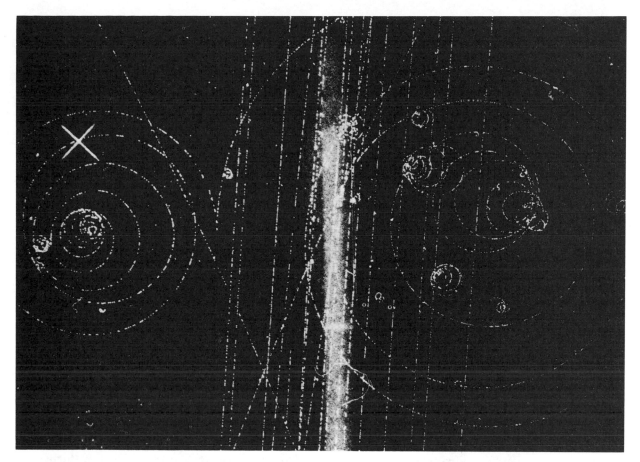

Figure 73 CREATION OF ANTIMATTER, here a positron, becomes visible in a bubble chamber where the trajectory of any electrically charged particle with an electric charge is marked by a trail of bubbles in liquid helium. The positron, whose clockwise spiraling path in the chamber magnetic field fills the right two-thirds of the photo, is the antiparticle of the electron, whose path is the smaller counterclockwise spiral. The electron-positron pair was created by a photon that is invisible because it has no charge. (Photo by Nicholas P. Samios.)

other and their mass is converted into energy. Hence a star made up of half matter and half antimatter would immediately disappear in a titanic explosion. The possibility remains, however, that matter and antimatter might coexist in the universe if each was confined to isolated regions separated by empty space.

One line of evidence for the preponderance of matter over antimatter is provided by cosmic rays, the high-energy particles that arrive from space. They seem invariably to be particles of matter such as protons and electrons and atomic nuclei made up of protons and neutrons; antiparticles are not observed. Although the origin of cosmic rays is not yet fully understood, they certainly come from

throughout the galaxy, and some of them may have a still more distant origin. It therefore seems established that the Milky Way consists entirely of matter, and it is only a little less certain that the group of galaxies of which the Milky Way is a member is also all matter.

Ascertaining that more distant galaxies are composed of matter is a more difficult problem. Merely looking at a galaxy offers no hint of whether it is made up of matter or antimatter. "Looking at" a galaxy implies the detection of photons, or quanta of electromagnetic radiation. The photons include not only those of visible light but also those of radio waves, X rays, gamma rays and so on. The problem is that the photon is its own antiparticle, and there is no way to distinguish a photon emitted by matter from one emitted by antimatter. As a result the light from an antimatter galaxy would be identical with that from a matter galaxy, even in the detailed structure of the spectrum. For example, the characteristic emission lines of the hydrogen atom would be duplicated exactly in emission lines of the antihydrogen atom.

There is one circumstance in which photon observations might indirectly reveal the presence of antimatter. If an antimatter galaxy were close to a matter galaxy, the boundary region between them would be the site of frequent particle-antiparticle annihilations. The energy of each such annihilation would eventually appear in the form of photons at gamma-ray wavelengths. The border region would therefore be a place where gamma radiation is copiously emitted. Astronomical sources of gamma radiation are known and are under investigation, but no source with the proper characteristics has been found. This argument is of no consequence, however, if empty space separates the matter from the antimatter. At best the failure to observe strong gamma emissions suggests that clusters of galaxies must consist entirely of matter or entirely of antimatter, not a mixture of the two. The clusters are pervaded by intergalactic gas, and any difference in composition within a cluster would give rise to gamma radiation.

In the future the question of whether any substantial aggregations of antimatter exist in the universe may be answered by the advent of telescopes that detect not photons but neutrinos. Unlike the photon, the neutrino has a distinguishable antiparticle. Neutrinos and antineutrinos would be emitted in different proportions by nuclear reactions in matter and antimatter. A star composed of matter radiates mainly neutrinos, whereas a star composed of antimatter would give rise chiefly to antineutrinos. The issue has not been settled yet by neutrino observations because building a neutrino telescope is a formidable project. Neutrinos have negligible mass and hardly interact at all with other matter; their detection is problematic.

For now at least, the prevailing opinion among astronomers and astrophysicists is that matter dominates over antimatter in the present universe. As I have suggested, the evidence in support of this view is not compelling, although there is a notable lack of evidence for the existence of antimatter. What ultimately seems decisive is the difficulty of imagining how matter and antimatter in the early universe

could have become segregated into distinct regions. It seems more likely they would have simply annihilated each other everywhere.

If the universe is now mostly matter, one is moved to ask how this asymmetry came about. One possibility is that the preference for matter was built in at the start, that the primordial material issuing from the big bang was predominantly matter. This hypothesis cannot be disproved, at least for now, but it is rather unsatisfying. Virtually any composition of the universe could be explained in the same way. Moreover, the primordial-imbalance hypothesis accords fundamental status to a set of initial conditions that have no apparent rationale; any number of alternatives seem equally plausible. If a theory consistent with established physical principles could be constructed in which the universe was initially symmetrical, it would be more appealing. It is just such a theory that is offered by the conjunction of cosmology and particle physics.

A crucial event in modern cosmology was the discovery in the 1920's by Edwin P. Hubble that distant galaxies are receding from the earth with speeds proportional to their distances. The recession of the galaxies implies that the entire universe is expanding. Extrapolating backward in time leads to the conclusion that roughly 10 billion years ago the material that now forms the galaxies emerged explosively from a highly compressed state. Indeed, following the backward evolution to its mathematical limit suggests that the entire universe was initially a dimensionless point.

At the instant of the big bang the density and the temperature of the universe were infinite. The temperature fell rapidly, but throughout the first minute it was greater than 10^{10} degrees Kelvin. Under those conditions any atoms that may have formed were immediately torn apart; even atomic nuclei could not survive but were decomposed into their constituent particles. In other words, the universe in its first moments was a hot plasma of free particles, many of which, such as the electrons and the protons, were electrically charged. Because charged particles in motion give off electromagnetic radiation, the early universe was rich in photons.

The expanding universe cooled much as an expanding gas cools, and by about three minutes after the big bang protons and neutrons began to combine to form the nuclei of helium atoms. The remaining unbound protons would eventually become hydrogen nuclei. (All the heavier elements, which are quite rare on a cosmic scale, have been built up out of hydrogen and helium in the cores of stars and in supernova explosions.) By making the simplest assumptions about the conditions in the early universe that are consistent with known physical laws, one can calculate that the ratio of helium to hydrogen was about one to three by weight. The value is in good agreement with the ratio estimated for the universe today. The success of this prediction is testimony to an understanding of what the universe was like a few minutes after its birth.

After roughly 10,000 years of expansion the universe was cool enough for the last of the free charged particles to be incorporated into atoms. Each atom is electrically neutral because it has equal numbers of positive and negative charges. Photons interact only weakly with neutral matter, and so from that time forward the matter and the electromagnetic radiation in the universe were essentially uncoupled. Since then the radiation has freely followed the expansion of the universe, cooling all the while. How can radiation cool, and how can it have a temperature in the first place? If the radiation is regarded as a gas of photons, then it cools by expansion, somewhat like a gas of material particles, as the average energy of the photons decreases. If the radiation is regarded as a wave, then the expansion of space brings an increase in the distance between any two successive wave crests. The longer wavelength corresponds to a smaller photon energy.

In 1964 it was discovered that microwave radiation is striking the earth evenly from all directions. The radiation corresponds to a photon gas that fills the universe to a density of about 300 photons per cubic centimeter. The temperature of the radiation is 2.7 degrees K., a value much reduced from the temperature of about 10,000 degrees at the time of decoupling. The presence of the radiation is further evidence that this theoretical reconstruction of the early universe is correct. Emboldened by these successes, one can attempt to extrapolate back to the earliest moments of the universe to see if the extreme conditions then prevailing might account for the present asymmetry between matter and antimatter.

In the first few seconds of the universe the particles of the hot primeval gas had an average energy that exceeds the capabilities of even the largest modern particle accelerators. Interactions of particles at those energies may have been qualitatively different from all those that can be observed now. Even if the events in the early universe differed in character from those accessible today, however, the laws of nature governing the events can be assumed to endure unchanged. What is needed, then, is a theory that will predict how particles act at very high energy on the basis of natural laws deduced from events at much lower energies.

Among the natural laws in question are conservation laws applied to quantum numbers. A quantum number is essentially a bookkeeping convenience, adopted as an aid to keeping track of the various properties of particles. For example, electric charge can be expressed as a quantum number: the proton is assigned a value of $+1$, the electron a value of -1 and the photon and all other neutral particles a value of zero. The conservation law that applies to electric charge states that the total charge quantum number cannot change in an interaction; the sum of all the charge quantum numbers after the event must be equal to the sum before the event.

It is important to note that the conservation of electric charge does not forbid a change in the number of charged particles. An electron and a positron can annihilate each other, diminishing the number of particles by two; the total charge, however, is zero both before the annihilation and after it. The opposite process, in which an electron and a positron are created out of pure energy, obeys the conservation law for the same reason. Indeed, any particle can be created or annihilated simultaneously with its antiparticle, and all quantum numbers will automatically be conserved.

A quantum number called baryon number is of notable interest in tracing the source of the cosmic asymmetry between matter and antimatter. The baryons are a large family of particles whose most familiar members are the proton and the neutron; as basic constituents of atomic nuclei, the baryons clearly have an important role in the structure of ordinary matter. The proton, the neutron and all the many related baryons are assigned a baryon number of $+1$. For the antiproton, the antineutron and other antibaryons the baryon number is -1. All other particles, including the pions, the muons, the neutrinos, the electron, the photon and their antiparticles, have a baryon number of zero.

The conservation of baryon number is the assertion that in any reaction the baryon number of all the particles in the initial state is equal to the baryon number of all the particles in the final state. Again the number of particles can change, as when a proton and an antiproton are created or annihilated as a pair, but the net baryon number remains unaltered. Suppose, for example, two protons (with a total baryon number of $+2$) collide at high energy. The final products might include four protons, a neutron, three antiprotons and a number of pions; adding up the baryon numbers shows that the total remains $+2$.

Electric charge is a quantity that is thought to be conserved under all circumstances. The absolute conservation of baryon number is less certain, and indeed there is now strong suspicion that the law is occasionally violated.

The most compelling evidence for the conservation of baryon number is the stability of the proton. As the least massive particle whose baryon number is $+1$, the proton cannot decay into any set of lighter particles without violating the conservation law. Detection of a proton decay would therefore constitute direct evidence that the law is not always enforced.

No one has yet seen a proton decay, and even crude calculations suggest that its lifetime is long. If protons decayed, for example, in human bone, the energy released would increase the incidence of cancer. On this basis the lifetime of the proton must be greater than 10^{16} years: If protons decayed on Jupiter, the energy would contribute to the luminosity of the planet. On this basis the lifetime is greater than 10^{18} years. Systematic experiments suggest that the lifetime is actually greater than 10^{29} years. In contrast, the age of the universe is only 10^{10} years. Evidently if the proton does decay, it is an exceedingly rare event. If the actual lifetime should turn out to be 10^{30} years, then in 100 tons of matter (a sample of 10^{31} protons) an average of 10 would decay in a year. The low rate suggests both the stringency of the law of conservation of baryon number and the difficulty of mounting experiments to search for violations. Several such experiments are nonetheless under way.

Saying that the universe has an excess of matter over antimatter is equivalent to saying that it has a positive baryon number. If the law of conservation of baryon number were absolute, the number would have been constant through the eons. There may have been more of both baryons and antibaryons once, but the number of baryons minus the number of antibaryons would have always been the same.

Consider the state of the universe when it was a hundredth of a second old and had a temperature of 10^{14} degrees K. For any given temperature there is an equilibrium mixture of different kinds of particles such that for each kind the number of particles being created by collisions or decays balances the number being destroyed. In the early universe, at 10^{14} degrees, the equilibrium mixture included about a billion protons and a billion antiprotons for every proton in the present universe. If the baryon number of the universe was the same then as it is now, the ratio of protons to antiprotons must have been roughly 1,000,000,001 to 1,000,000,000, and so the asymmetry would have been scarcely noticeable.

Later almost all the protons were annihilated by encounters with antiprotons. Only the conservation of baryon number forestalled a total annihilation of all baryons and antibaryons. In this view all the present protons, and therefore all the present galaxies, stars, planets and sentient beings, are the residue of a one-part-in-a-billion imbalance. It is the small imbalance, the early manifestation of the cosmic asymmetry between matter and antimatter, that stands in need of explanation. Once the excess of matter has been established the subsequent evolution of the universe is comparatively straightforward; the source of the original asymmetry is a deeper mystery. In particular, if the universe evolved from an initial state that was fully symmetrical between matter and antimatter (a state having a baryon number of zero) into an asymmetrical state in which the baryon number is greater than zero and protons outnumber antiprotons, then the conservation of baryon number must have been violated at some stage.

The first indication that the conservation of baryon number cannot be exact came from a distantly related field of inquiry: the theory of the black hole. A mathematical analysis demonstrated that the only properties of a black hole measurable by an outside observer are its mass, its angular momentum and its electric charge. Notably absent from the list is the baryon number. Hence a black hole created by the collapse of a star would be indistinguishable from one created by the collapse of an antistar with the same mass, angular momentum and charge. Yet the baryon number for the star is positive, whereas the number for the antistar is negative. Clearly there is no way to assign a baryon number to a black hole and be certain that the baryon number of the universe is conserved.

The putative violation of the conservation law by black holes suggests that a similar mechanism on a microscopic scale might lead to proton decay. In this hypothetical process a proton is absorbed by a virtual black hole: a minute, short-lived fluctuation in the geometry of space-time, which in principle could arise anywhere and at any time. The virtual black hole promptly decays into a positron and a gamma ray. In these particles the mass or energy of the proton reappears and so does its positive electric charge; its baryon number, however, is irretrievably lost. Although the details of the hypothetical process are uncertain, estimates suggest that it implies a lifetime for the proton on the order of 10^{40} years. If the conservation of baryon number is violated in this way, the violation is feeble indeed.

A second indication that the conservation of baryon number is only approximate is slightly less exotic and also more powerful in its effect on the lifetime of the proton. This second possible mechanism is an outcome of revolutionary developments in the theories describing interactions among elementary particles. To be specific, it is an outcome of the understanding, achieved only in the past decade, that the "'strong" force responsible for holding together the nuclei of atoms and the "weak" force responsible for most radioactive decays are quite similar to electromagnetism.

How could improved understanding of these forces, which do not violate the conservation of baryon number, lead to theories predicting such a violation? A more detailed discussion of the forces must precede the explanation.

Of the three forces only electromagnetism is routinely evident in the macroscopic world that people perceive directly. The electromagnetic force acts only between particles that have an electric charge; the interaction can be described as the exchange of a third particle, namely a photon. The photon is said to be a vector particle, a designation given to any particle whose spin angular momentum, when measured in fundamental units, is equal to 1. Perhaps the most fundamental characteristic of electromagnetism is that it can be described by a gauge-invariant theory. In a theory of this kind the origin of the force is related to a conservation law, in this case the conservation of electric charge. The coupling of vector particles to a conserved charge is characteristic of gauge theories (see Chapter 5, Figure 38).

In all these respects the strong interaction is similar. The force arises from a gauge theory, and a strong interaction can be described as the exchange of a vector particle by two other particles that have a certain kind of charge. The vector particle is not

the photon, however, but a hypothetical entity called a gluon, and the charge is not electric charge but a property called color. The color charge of course has nothing to do with color in the ordinary sense. The word charge in this context is less fanciful. The word is apt because color charge plays much the same role in the strong interaction as electric charge does in the electromagnetic interaction.

One difference between electromagnetism and the strong interaction is that electromagnetism has only one kind of charge, whereas in the strong interaction there are three, labeled R, G and B for red, green and blue. The colors are carried by the funda-

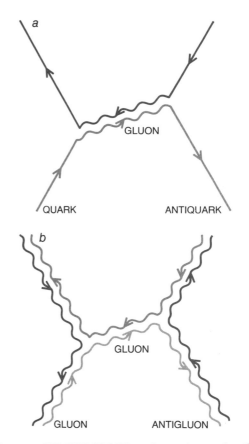

Figure 74 STRONG FORCE can be represented as the exchange of a gluon between two particles that have color. In *a* the particles are quarks; the one on the left is blue, the one on the right is antiblue. The strong interaction changes the trajectory and the color of each quark. To conserve color throughout the interaction the gluon must have both a color and an anticolor; as a result the gluons themselves are subject to the strong interaction. Gluon and antigluon scattering is shown in *b*.

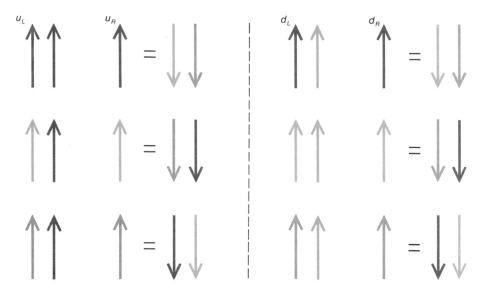

Figure 75 FAMILY OF 12 QUARKS. Each quark has four salient properties: type (e.g., *u* or *d*), color charge with respect to the strong interaction (red, green or blue), spin and for left-handed particles weak charge (purple for *u*, orange for *d*). In the mathematical formalism of the strong interaction a quark with a given color is equivalent to a quark without that color but with two anticolors (shown for right-handed quarks).

mental constituents of all strongly interacting particles: the quarks. Each quark has a single color, denoted by an assignment of the three color quantum numbers. For red quarks R equals $+1$ whereas G equals 0 and B equals 0. Similarly, for green quarks G equals $+1$ and for blue quarks B equals $+1$ and the other color quantum numbers are zero. Eight kinds of gluon are required by the theory. Six kinds change a quark of one color into a quark of a different color in all possible ways, namely red into green, red into blue, green into red, green into blue, blue into red and blue into green. The other two gluons resemble the photon in that they carry a force between "charged" particles but do not alter the charge.

In addition to color quarks have three other properties. Each quark is called up (*u*) or down (*d*). The *u* quarks have an electric charge of $+\frac{2}{3}$, the *d* quarks an electric charge of $-\frac{1}{3}$. Each quark also has a spin whose axis is aligned with the particle's direction of motion (subscript R for right-handed particles), or is opposite to that direction (subscript L for left-handed particles). Finally, the left-handed quarks have a color with respect to the force in nature called the weak interaction, which mediates most radioactive decays (purple for *u*, orange for *d*).

A property of color charges is that they can cancel one another. For example, the combination of one

red, one green and one blue quark is a colorless composite particle, to which gluons do not couple. (Similarly, particles with opposite electric charges can combine to form a neutral composite.) It is only such colorless combinations of quarks that seem to appear in nature. All baryons consist of three quarks, one quark in each of the three colors. The mesons, which make up another category of strongly interacting particles, each consist of a quark and an antiquark.

A second difference between the strong interaction and electromagnetism is that the gluons themselves are charged, whereas the photon is not. For example, the gluon that is absorbed by a red quark and transforms it into a green quark has R equal to -1, G equal to $+1$ and B equal to 0; with this combination of colors and anticolors color charge is conserved throughout the interaction. Since gluons couple to colored particles and since gluons themselves are colored, gluons couple to one another. In contrast, the photon is electrically neutral and does not couple to other photons. The difference has a profound dynamical consequence: at short distances the strong interaction loses strength. Quarks bind only feebly when they are close together, but their binding becomes quite powerful when they are somewhat farther apart. (In the present context a long distance is 10^{-13} centimeter.)

This paradoxical force law explains a great deal. It has been known since the mid-1960's that the properties of strongly interacting particles could be accounted for by the quark model, but no one has ever observed an isolated quark. Furthermore, the utility of treating a strongly interacting particle as a composite of quarks rests on an approximation in which the quarks are essentially noninteracting particles inside a communal "bag." It was puzzling that strongly interacting particles such as quarks could successfully be described as noninteracting. The notion that the strength of the strong interaction among quarks decreases when the quarks are close together neatly explains why the quarks inside a "bag" interact only feebly with one another and yet cannot be pulled far apart. It may be impossible to isolate a quark. The gauge theory of the strong interaction that underlies the quark model leads to many experimental predictions, which so far have proved very successful. The theory is gaining almost universal acceptance.

The weak interaction can be described in much the same way as the electromagnetic and the strong interactions, but it has a few twists of its own. First, there are two kinds of charge, analogous to the three color charges of the strong interaction. I shall call them P and O, for the colors purple and orange. Three vector particles, called W^+, W^- and Z, mediate the interaction. These particles have large masses, unlike the photon and the gluons, which are massless. A particle with a large mass can arise spontaneously only as a short-lived fluctuation; if it is short-lived, it cannot go far, and as a result the weak interaction has a very short range. A more surprising characteristic of the weak force is that it acts only on particles with certain geometric properties. Quarks, electrons, neutrinos and a few other particles can be classified as right-handed or left-handed according to the relative orientation of their spin angular momentum and their linear motion. A

right-handed particle has its spin axis pointing parallel to its direction of motion, a left-handed particle antiparallel. The weak interaction affects only left-handed particles and right-handed antiparticles. In sum, the strong and the weak interactions require five kinds of color charge (red, green and blue for the strong and purple and orange for the weak), along with vector particles that transmute some of these colors.

In the theories I have outlined here the strong force is a mechanism for changing the red, green and blue colors of quarks. The weak force works similar changes on the purple and orange color quantum numbers of particles. If these theories are to be truly unified, one would expect some additional force to transform the strong colors into weak colors and vice versa. In addition to being aesthetically attractive, a scheme that incorporates such a new force accommodates all known particles quite neatly. Moreover, it makes definite predictions. For example, it predicts the mass of the W.

It is by postulating a new force that the unifying theories compromise the conservation of baryon number and allow the proton to decay. New color-changing vector particles are introduced as bridges between particles with strong color, such as the quarks composing a proton, and particles with only weak color, whose baryon number is zero. I shall designate such vector particles X. The unifying theory predicts that the X has a mass that is 10^{15} times the mass of the proton (and is roughly comparable to the mass of a flea), compressed into a volume only 10^{-27} centimeter across. Because the X particle is so massive, its spontaneous creation is extremely rare. Accordingly it is estimated that the mean lifetime of the proton is long but not infinite; the lifetime should be on the order of 10^{31} years.

To be sure, a lifetime of 10^{31} years implies that in the universe today the violation of the conservation of baryon number is slight. As I have noted, however, the matter-antimatter asymmetry observed

Figure 76 WEAKLY INTERACTING PARTICLES include the left-handed neutrino; its antiparticle, the right-handed antineutrino; the left-handed electron, and its antiparticle, the right-handed positron. These particles do not interact strongly and so they are shown without strong color charges. The right-handed electron (*right*) is given two weak color charges, a configuration equivalent to a state with no weak color charge (*black line without arrow*).

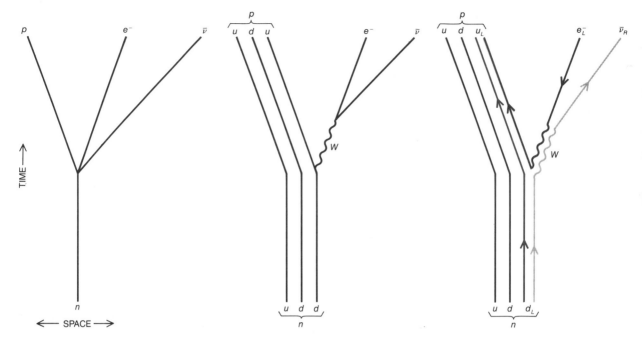

TIME →

← SPACE →

Figure 77 DECAY OF THE NEUTRON by a weak interaction averages 15 minutes. In the broadest view (*left*) the decay transforms the neutron (*n*) into a proton (*p*), an electron (*e⁻*) and an antineutrino (*v̄*). In a finer analysis (*center*) the neutron consists of three quarks and only a *d* quark, arbitrarily placed at the right, is affected by the decay. A left-handed *d* quark (*right*) decays into the left-handed *u* quark, the electron and the antineutrino. The orange charge of the d_L quark is transformed by a short-lived (or virtual) particle *W*, which is orange and antipurple. Only weak colors are changed.

today corresponds to merely a one-part-in-a-billion asymmetry in the early universe. Moreover, a mode of decay that requires the creation of an unstable heavy particle may well have been commoner in the earliest moments of the universe, when heavy particles could be freely created by ultrahigh-energy collisions.

I turn now to the idea that physical laws are indifferent to the distinction between matter and antimatter. The history of the idea is a series of upset expectations. Until the mid-1950's it was generally thought the laws of physics would remain unchanged if experiments were repeated in a mirror-reflected world. In other words, it was thought no absolute distinction could be made between left and right. A variety of experiments then revealed, however, that mirror-reflection symmetry is badly broken by the weak interactions. An example is provided by the decay of the muon into an electron, a neutrino and an antineutrino. In more than 999 decays in 1,000 the electron is found to be left-handed: its spin axis points in the direction opposite to its direction of motion. Thus the decay of the muon furnishes an absolute standard of left v. right.

Theorists next proposed a more comprehensive symmetry that seemed to be respected by all interactions. This second hypothesis was that the laws of physics would be unchanged by the mirror reflection of an experiment if at the same time all the particles in the experiment were replaced by their antiparticles. The symmetry is called CP for charge conjugation and parity, or mirror reflection. CP symmetry predicts that in the decay of the antimuon a positron should emerge instead of an electron and the positron should almost always be right-handed. In the case of muon decay exact CP symmetry is observed.

If CP symmetry were absolute, a preponderance of matter or of antimatter could not evolve from a primordial equality between the two. For every process that creates a particle, an equally likely mirror process would create the antiparticle.

The concept of absolute CP symmetry survived

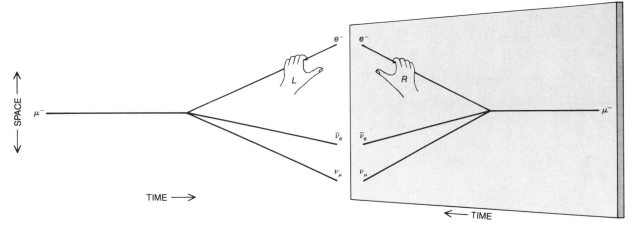

Figure 78 VIOLATION OF PARITY SYMMETRY. Parity, or P, conservation holds that processes remain invariant when they are transformed by a mirror image. The process shown is the decay of a muon (μ) into an electron (e^-), an electron-type antineutrino ($\bar{\nu}_e$) and a muon-type neutrino (ν_μ). The electron is left-handed. In the mirror reflection of the decay the electron is right-handed. In reality parity symmetry is broken in this process: left-handed electrons appear more than 1,000 times as often as right-handed ones.

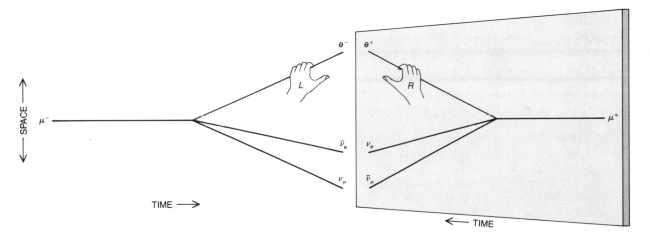

Figure 79 CP SYMMETERY proposed a symmetry that might be observed even if parity symmetry is violated, asserting that the symmetry broken by mirror reflection could be restored by replacing all particles with their anti-particles. In muon decay CP symmetry holds true: decay at the left and right appear to be equally common.

for about seven years. Then it was observed that the long-lived neutral K meson, which is its own anti-particle, decays more often into a negative pion, a positron and a neutrino than it does into a positive pion, an electron and an antineutrino. If CP were an absolute symmetry, the two decay modes would have to be equally likely. No violation of CP symmetry has been found except in K-meson decay, but

such violations might have a more prominent role in nature at ultrahigh energies.

The developments I have described suggest that both the permanence of certain particles, as formalized in the law of conservation of baryon number, and the indifference of physical laws to the distinction between matter and antimatter, as for-

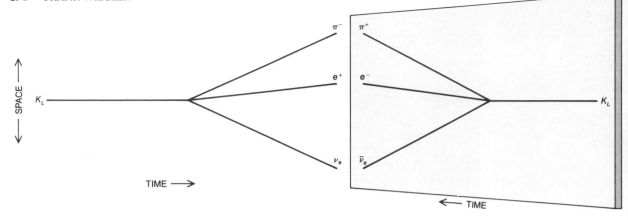

Figure 80 CP VIOLATION was demonstrated in the decay of the long-lived neutral K meson (K_L). The decay of this particle into π^-, e^+ and v_e is commoner than decay into the antiparticles π^+, e^- and $\overline{v}_e$. (The long-lived neutral K meson is its own antiparticle.) If CP symmetry were never broken, the ratio of baryons to antibaryons would be fixed and no asymmetry could develop between matter and antimatter.

malized in the principle of CP symmetry, are not exact but only approximate. It is true the principles hold quite accurately today, but this may not have been the case in the very early universe. Indeed, given even a small violation of these principles one can construct a specific chain of events leading from a universe in an initial state of symmetry between matter and antimatter to a universe with a preponderance of matter over antimatter.

The chain of reasoning begins with the observation that the temperature of the universe has been falling steadily since the big bang. The higher the temperature, the higher the average speed and energy of the particles that make up the universe, and hence the greater the energy available in a collision for the creation of other particles. At a temperature greater than 10^{28} degrees K. the typical energy of a particle was comparable to the rest-mass energy of an X particle. Until about 10^{-35} second after the big bang the universe had such a temperature, and so one can propose that it had a great density of X particles.

As the universe expanded and cooled, the probability of creating an X particle declined rapidly; meanwhile the existing particles were rapidly decaying. Suppose the decays did not conserve baryon number. An X particle might then decay into any of several final states with differing total baryon number. The average might be, say, $+\frac{2}{3}$. If the universe had equal amounts of matter and antimatter before it was 10^{-35} second old, it would include equal numbers of X's and $\overline{X}$'s, where the $\overline{X}$ is the antipar-

ticle of an X. It might seem, therefore, that every decay mode of an X would be counterbalanced by the decay of an $\overline{X}$, which would yield particles with an average baryon number of $-\frac{2}{3}$. In that case the total baryon number of the universe would remain zero at all times. Actually, since CP symmetry may not have been observed exactly in the decay of the X and the $\overline{X}$, one cannot conclude that the two decay sequences always yielded symmetrically opposite sets of particles. The $\overline{X}$ might give rise to particles whose average baryon number was not $-\frac{2}{3}$ but rather, say, $-\frac{1}{3}$.

In this way a universe that had equal numbers of X and $\overline{X}$ particles would have evolved into a universe with a positive baryon number and a corresponding preponderance of matter. It could have been a universe, for example, with a one-part-in-a-billion imbalance favoring matter. After the first 10^{-35} second or so the temperature and the typical energy per particle throughout the universe would fall below the threshold for the creation of an X and an $\overline{X}$. The processes that violate baryon number would then become insignificant, and the preponderance of matter over antimatter would be frozen in. The universe would still have many more baryons and antibaryons than it has now, but most of them would eventually annihilate one another, leaving the residue of matter observed today.

Several aspects of this argument are highly speculative, and the explanation of the cosmic asymmetry between matter and antimatter may seem more mythical than scientific. To an extent that is unavoidable, since the extreme conditions of the early universe cannot be reproduced in a laboratory. What distinguishes scientific speculation from myth is its logical consistency and the amenability of at least some of its elements to experimental test. I have described how the inner logic of particle physics leads to unified theories in which baryon number is not conserved, and I have noted that future developments both in neutrino astronomy and in the search for the decay of the proton will test the theories. If these difficult experiments give results consistent with theoretical expectations, they will bring much closer the scientific understanding of a mysterious asymmetry. Even now, calculations carried out in accordance with the unified theories suggest that the average density of matter in the universe today is consistent with the primordial course of events the unified theories imply. Because of uncertainties about the mechanisms of CP violation it is difficult to make the calculations precise, but the qualitative picture is satisfying.

A further question remains. I have described how the universe could have begun with symmetry between matter and antimatter and then have grown asymmetrical. Why was the universe symmetrical in the beginning?

At one level this question can be answered statistically. Even if interactions that violate baryon number were frequent in the early universe, the most likely universal condition, which would be attained at equilibrium before 10^{-35} second, is one in which the amount of matter equals the amount of antimatter. Unified theories therefore enforce initial symmetry automatically; it need not be postulated separately. After 10^{-35} second the decay rates of the X's and $\overline{X}$'s would have been slow compared with the expansion and cooling rate of the universe. Under the condition equilibrium could no longer be attained.

At a deeper level I do not find this explanation fully satisfying. It fails to explain why the universe should have begun in an explosive event. It also

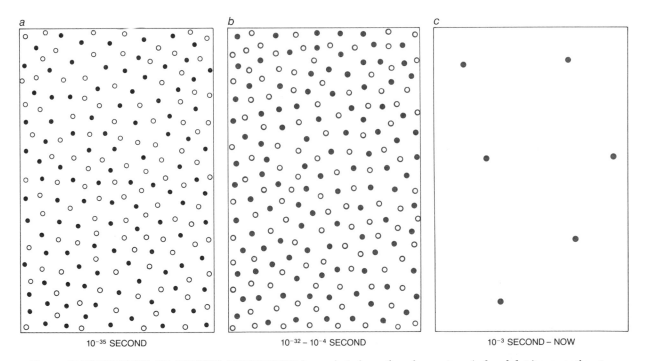

10⁻³⁵ SECOND   10⁻³² – 10⁻⁴ SECOND   10⁻³ SECOND – NOW

Figure 81 EVOLUTION OF COSMIC AYSMMETRY between matter and antimatter. Panel *a*, the universe 10^{-35} second after the big bang, shows equal quantities of X particles (*dots*) and their antiparticles, $\overline{X}$'s (*open circles*). In panel *b* the X's and $\overline{X}$'s have decayed leaving a slight imbalance favoring protons (*colored dots*) over antiprotons (*open circles*). In panel *c* each encounter of a proton and an antiproton has caused the annihilation of both particles, and only the excess protons survive (six in the drawing, one in a billion actually).

fails to explain why the universe is symmetrical in several other ways: it is electrically neutral on the average and it seems to have no net angular momentum.

I shall now describe an idea that may lead to an understanding of these questions. It is by no means well established, but it does suggest a program of research. Indeed, it was the original motivation for my own work on the matter-antimatter asymmetry.

Modern theories of the interactions among elementary particles suggest that the universe can exist in different phases that are analogous in a way to the liquid and solid phases of water. In the various phases the properties of matter are different; for example, a certain particle might be massless in one phase but massive in another. The laws of physics are more symmetrical in some phases than they are in others, just as liquid water is more symmetrical than ice, in which the crystal lattice distinguishes certain positions and directions in space.

In these theories the most symmetrical phase of the universe generally turns out to be unstable. One can speculate that the universe began in the most symmetrical state possible and that in such a state no matter existed; the universe was a vacuum. A second state was available, and in it matter existed. The second state had slightly less symmetry, but it was also lower in energy. Eventually a patch of the less symmetrical phase appeared and grew rapidly. The energy released by the transition found form in the creation of particles. This event might be identified with the big bang. The electrical neutrality of the universe of particles would then be guaranteed, because the universe lacking matter had been electrically neutral. The lack of rotation in the universe of matter could be understood as being among the conditions most favorable for the phase change and the subsequent growth, with all that the growth implied, including the cosmic asymmetry between matter and antimatter. The answer to the ancient question "Why is there something rather than nothing?" would then be that "nothing" is unstable.

POSTSCRIPT

Since my article was written, there have been two important experimental developments. New experiments at high-energy accelerators have dramatically supported and strengthened our faith in what is now called the "standard model" of particle interactions. At the CERN proton-antiproton collider, the $W^\pm$ and Z^0 bosons were produced and convinc-

ingly identified, and the properties of quark and gluon jets predicted by QCD were confirmed in accurate qualitative detail. The standard model ascribes the strong, electromagnetic and weak interactions to exchange of various gauge bosons — respectively, the color gluons of QCD, the photon, and the W and Z particles — which are naturally thought of as either responding to, or transforming, five different "color" charges. These separate theories fairly cry out for unification into a single theory that allows all possible transformations among the five colors. Theories embodying this unification invariably lead to violation of the law of baryon-number conservation, an essential requirement for a theoretical explanation of the cosmic matter-antimatter asymmetry.

So much for the good news. The bad news is that despite heroic efforts the one manifestation of unification that might have been accessible in terrestrial laboratories — decay of the proton due to baryon number nonconservation — has not been seen. In fact, the proton lifetime almost certainly exceeds 10^{32} years. More sensitive experiments are mainly limited by the sheer size of the necessary detectors. If in several years not a single proton decays within 100 kilotons of material, both the patience of experimenters and their ability to monitor gigantic chunks of matter become severely strained.

The simplest unified models predict proton lifetimes of about 10^{30} years, so they are now pretty clearly excluded. But while violation of the law of baryon number conservation, and hence the possibility of proton decay, is a robust consequence of unification, the exact value of the predicted lifetime depends sensitively on details of how the basic idea is implemented in a concrete model. So we have the unsatisfactory situation that the theoretical idea of unification seems more attractive than ever, but there is no good idea for testing it experimentally.

How will we escape from this impasse? There are a few ideas for other experiments that could in principle reveal some sign of the physics associated with unification. Detection of cosmic axions or of neutrino masses, are genuine possibilities. These possibilities may already be physical realities as the mysterious cosmic "dark matter" could well consist of axions or massive neutrinos, and the absence of expected neutrinos from the sun may possibly indicate changes in their properties (oscillations) resulting from their having mass.

Also, some of the exotic objects associated with

unification—magnetic monopoles or cosmic strings—might have been produced in early states of the big bang and persisted to the present day. Experimental searches for such exotics are being rigorously pursued. On the other hand, one of the main advantages of the inflationary universe idea (discussed in Chapter 11) is precisely to get rid of these objects, which would otherwise excessively clutter up the universe.

Finally, it is possible that a unified theory will emerge that is so complete and compelling that its consequences, including violation of the law of baryon-number conservation, will be accepted even without direct tests. Such a theory would, of course, have to have other consequences that can be tested directly. It might, for example, predict the ratio of electron to muon mass, the observed magnitude of CP symmetry violation or the existence and properties of hitherto unobserved particles at future accelerators. At present many physicists attach high hopes to superstring theories, but as yet no concrete predictions have been extracted from these theories.

In summary, the attractive speculations discussed in my article remain attractive speculations. There is considerable hope, but no certainty, that within the present millenium they will become more than that.

The Inflationary Universe

A new theory of cosmology suggests that the observable universe is embedded in a much larger region of space that had an extraordinary growth spurt a fraction of a second after the primordial big bang.

. . .

Alan H. Guth and Paul J. Steinhardt

I n the past few years certain flaws in the standard big-bang theory of cosmology have led to the development of a new model of the very early history of the universe. The model, known as the inflationary universe, agrees precisely with the generally accepted description of the observed universe for all times after the first 10^{-30} second. For this first fraction of a second, however, the story is dramatically different. According to the inflationary model, the universe had a brief period of extraordinarily rapid inflation, or expansion, during which its diameter increased by a factor perhaps 10^{50} times larger than had been thought. In the course of this stupendous growth spurt all the matter and energy in the universe could have been created from virtually nothing. The inflationary process also has important implications for the present universe. If the new model is correct, the observed universe is only a very small fraction of the entire universe.

The inflationary model has many features in common with the standard big-bang model. In both models the universe began between 10 and 15 billion years ago as a primeval fireball of extreme density and temperature, and it has been expanding and cooling ever since. This picture has been successful in explaining many aspects of the observed universe, including the red-shifting of the light from distant galaxies, the cosmic microwave background radiation and the primordial abundances of the lightest elements. All these predictions have to do only with the events that presumably took place after the first second, when the two models coincide.

Until recently there were few serious attempts to describe the universe during its first second. The temperature in this period is thought to have been higher than 10 billion degrees Kelvin, and little was known about the properties of matter under such conditions. Relying on recent developments in the physics of elementary particles, however, cosmologists are now attempting to understand the history of the universe back to 10^{-45} second after its beginning. (At even earlier times the energy density would have been so great that Einstein's general theory of relativity would have to be replaced by a quantum theory of gravity, which so far does not exist.) When the standard big-bang model is extended to these earlier times, various problems arise. First, it becomes clear that the model requires a number of stringent, unexplained assumptions about the initial conditions of the universe. In addition most of the new theories of elementary particles imply that the standard model would lead to a tremendous overproduction of the exotic particles called magnetic monopoles (each of which corresponds to an isolated north or south magnetic pole).

The inflationary universe was invented to overcome these problems. The equations that describe the period of inflation have a very attractive feature: from almost any initial conditions the universe evolves to precisely the state that had to be assumed as the initial one in the standard model. Moreover, the predicted density of magnetic monopoles becomes small enough to be consistent with observations. In the context of the recent developments in elementary-particle theory the inflationary model seems to be a natural solution to many of the problems of the standard big-bang picture.

The standard big-bang model is based on several assumptions. First; it is assumed that the fundamental laws of physics do not change with time and that the effects of gravitation are correctly described by Einstein's general theory of relativity. It is also assumed that the early universe was filled with an almost perfectly uniform, expanding, intensely hot gas of elementary particles in thermal equilibrium. The gas filled all of space, and the gas and space expanded together at the same rate. When they are averaged over large regions, the densities of matter and energy have remained nearly uniform from place to place as the universe has evolved. It is further assumed that any changes in the state of the matter and the radiation have been so smooth that they have had a negligible effect on the thermodynamic history of the universe. The violation of the last assumption is a key to the inflationary-universe model.

The big-bang model leads to three important, experimentally testable predictions. First, the model predicts that as the universe expands, galaxies recede from one another with a velocity proportional to the distance between them. In the 1920's Edwin P. Hubble inferred just such an expansion law from his study of the red shifts of distant galaxies. Second, the big-bang model predicts that there should be a background of microwave radiation bathing the universe as a remnant of the intense heat of its origin. The universe became transparent to this radiation several hundred thousand years after the big bang. Ever since then the matter has been clumping into stars, galaxies and the like, but the radiation has simply continued to expand and red-shift, and in effect to cool. In 1964 Arno A. Penzias and Robert W. Wilson of the Bell Telephone Laboratories discovered a background of microwave radiation

received uniformly from all directions with an effective temperature of about three degrees K. Third, the model leads to successful predictions of the formation of light atomic nuclei from protons and neutrons during the first minutes after the big bang. Successful predictions can be obtained in this way for the abundance of helium 4, deuterium, helium 3 and lithium 7. (Heavier nuclei are thought to have been produced much later in the interior of stars.)

Unlike the successes of the big-bang model, all of which pertain to events a second or more after the big bang, the problems all concern times when the universe was much less than a second old. One set of problems has to do with the special conditions the model requires as the universe emerged from the big bang.

The first problem is the difficulty of explaining the large-scale uniformity of the observed universe. The large-scale uniformity is most evident in the microwave background radiation, which is known to be uniform in temperature to about one part in 10,000. In the standard model the universe evolves much too quickly to allow this uniformity to be achieved by the usual processes whereby a system approaches thermal equilibrium. The reason is that no information or physical process can propagate faster than a light signal. At any given time there is a maximum distance, known as the horizon distance, that a light signal could have traveled since the beginning of the universe (see Chapter 3, Figure 18). In the standard model the sources of the microwave background radiation observed from opposite directions in the sky were separated from each other by more than 90 times the horizon distance when the radiation was emitted. Since the regions could not have communicated, it is difficult to see how they could have evolved conditions so nearly identical.

The puzzle of explaining why the universe appears to be uniform over distances that are large compared with the horizon distance is known as the horizon problem. It is not a genuine inconsistency of the standard model; if the uniformity is assumed in the initial conditions, the universe will evolve uniformly. The problem is that one of the most salient features of the observed universe—its large-scale uniformity—cannot be explained by the standard model; it must be assumed as an initial condition.

Even with the assumption of large-scale uniformity, the standard big-bang model requires yet another assumption to explain the nonuniformity observed on smaller scales. To account for the clumping of matter into galaxies, clusters of galaxies, superclusters of clusters and so on, a spectrum of primordial inhomogeneities must be assumed as part of the initial conditions. The fact that the spectrum of inhomogeneities has no explanation is a drawback in itself, but the problem becomes even more pronounced when the model is extended back to 10^{-45} second after the big bang. The incipient clumps of matter develop rapidly with time as a result of their gravitational self-attraction, and so a model that begins at a very early time must begin with very small inhomogeneities. To begin at 10^{-45} second the matter must start in a peculiar state of extraordinary but not quite perfect uniformity. A normal gas in thermal equilibrium would be far too inhomogeneous, owing to the random motion of particles. This peculiarity of the initial state of matter required by the standard model is called the smoothness problem.

Another subtle problem of the standard model concerns the energy density of the universe. According to general relativity, the space of the universe can in principle be curved, and the nature of the curvature depends on the energy density. If the energy density exceeds a certain critical value, which depends on the expansion rate, the universe is said to be closed: space curves back on itself to form a finite volume with no boundary. (A familiar analogy is the surface of a sphere, which is finite in area and has no boundary.) If the energy density is less than the critical density, the universe is open: space curves but does not turn back on itself, and the volume is infinite. If the energy density is just equal to the critical density, the universe is flat: space is described by the familiar Euclidean geometry (again with infinite volume).

The ratio of the energy density of the universe to the critical density is a quantity cosmologists designate by the Greek letter Ω (omega). The value $\Omega = 1$ (corresponding to a flat universe) represents a state of unstable equilibrium. If Ω was ever exactly equal to 1, it would remain exactly equal to 1 forever. If Ω differed slightly from 1 an instant after the big bang, however, the deviation from 1 would grow rapidly with time. Given this instability, it is surprising that Ω is measured today as being between .1 and 2. (Cosmologists are still not sure whether the universe is open, closed or flat.) In order for Ω to be in this rather narrow range today, its value a second after the big bang had to equal 1

to within one part in 10^{15}. The standard model offers no explanation of why Ω began so close to 1 but merely assumes the fact as an initial condition. This shortcoming of the standard model, called the flatness problem, was first pointed out in 1979 by Robert H. Dicke and P. James E. Peebles of Princeton University.

The successes and drawbacks of the big-bang model we have considered so far involve cosmology, astrophysics and nuclear physics. As the big-bang model is traced backward in time, however, one reaches an epoch for which these branches of physics are no longer adequate. In this epoch all matter is decomposed into its elementary-particle constituents. In an attempt to understand this epoch cosmologists have made use of recent progress in the theory of elementary particles. Indeed, one of the important developments of the past decade has been the fusing of interests in particle physics, astrophysics and cosmology. The result for the big-bang model appears to be at least one more success and at least one more failure.

Perhaps the most important development in the theory of elementary particles over the past decade has been the notion of grand unified theories, the prototype of which was proposed in 1974 by Howard M. Georgi and Sheldon Lee Glashow of Harvard University. The theories are difficult to verify experimentally because their most distinctive predictions apply to energies far higher than those that can be reached with particle accelerators. Nevertheless, the theories have some experimental support, and they unify the understanding of elementary-particle interactions so elegantly that many physicists find them extremely attractive.

The basic idea of a grand unified theory is that what were perceived to be three independent forces —the strong, the weak and the electromagnetic— are actually parts of a single unified force. In the theory a symmetry relates one force to another. Since experimentally the forces are very different in strength and character, the theory is constructed so that the symmetry is spontaneously broken in the present universe.

A spontaneously broken symmetry is one that is present in the underlying theory describing a system but is hidden in the equilibrium state of the system. For example, a liquid described by physical laws that are rotationally symmetric is itself rotationally symmetric: the distribution of molecules

TYPE OF UNIVERSE	RATIO OF ENERGY DENSITY TO CRITICAL DENSITY (Ω)	SPATIAL GEOMETRY	VOLUME	TEMPORAL EVOLUTION
CLOSED	>1	POSITIVE CURVATURE (SPHERICAL)	FINITE	EXPANDS AND RECOLLAPSES
OPEN	<1	NEGATIVE CURVATURE (HYPERBOLIC)	INFINITE	EXPANDS FOREVER
FLAT	1	ZERO CURVATURE (EUCLIDEAN)	INFINITE	EXPANDS FOREVER, BUT EXPANSION RATE APPROACHES ZERO

Figure 82 THREE TYPES OF UNIVERSE—closed, open and flat—can arise from the standard big-bang model. The distinction between the different geometries depends on the quantity designated Ω, the ratio of energy density of the universe to some critical density, whose value depends on the rate of expansion of the universe.

looks the same no matter how the liquid is turned. When the liquid freezes into a crystal, however, the atoms arrange themselves along crystallographic axes and the rotational symmetry is broken. One would expect that if the temperature of a system in a broken-symmetry state were raised, it could undergo a kind of phase transition to a state in which the symmetry is restored, just as a crystal can melt into a liquid. Grand unified theories predict such a transition at a critical temperature of roughly 10^{27} degrees.

One novel property of the grand unified theories has to do with the particles called baryons, a class whose most important members are the proton and the neutron. In all physical processes observed up to now the number of baryons minus the number of antibaryons does not change; in the language of particle physics the total baryon number of the system is said to be conserved. A consequence of such a conservation law is that the proton must be absolutely stable; because it is the lightest baryon, it cannot decay into another particle without changing the total baryon number. Experimentally the lifetime of the proton is known to exceed 10^{31} years.

Grand unified theories imply that baryon number is not exactly conserved. At low temperature, in the broken-symmetry phase, the conservation law is an excellent approximation, and the observed limit on the proton lifetime is consistent with at least many versions of grand unified theories. At high temperature, however, processes that change the baryon number of a system of particles are expected to be quite common.

One direct result of combining the big-bang model with grand unified theories is the successful prediction of the asymmetry of matter and antimatter in the universe. It is thought that all the stars, galaxies and dust observed in the universe are in the form of matter rather than antimatter; their nuclear particles are baryons rather than antibaryons. It follows that the total baryon number of the observed universe is about 10^{78}. Before the advent of grand unified theories, when baryon number was thought to be conserved, this net baryon number had to be postulated as yet another initial condition of the universe. When grand unified theories and the big-bang picture are combined, however, the observed excess of matter over antimatter can be produced naturally by elementary-particle interactions at temperatures just below the critical temperature of the phase transition. Calculations in the grand unified theories depend on too many arbitrary parameters for a quantitative prediction, but the observed matter-antimatter asymmetry can be produced with a reasonable choice of values for the parameters.

A serious problem that results from combining grand unified theories with the big-bang picture is that a large number of defects are generally formed during the transition from the symmetric phase to the broken-symmetry phase. The defects are created when regions of symmetric phase undergo a transition to different broken-symmetry states. In an analogous situation, when a liquid crystallizes, different regions may begin to crystallize with different orientations of the crystallographic axes. The domains of different crystal orientation grow and coalesce, and it is energetically favorable for them to smooth the misalignment along their boundaries. The smoothing is often imperfect, however, and localized defects remain.

In the grand unified theories there are serious cosmological problems associated with pointlike defects, which correspond to magnetic monopoles, and surfacelike defects, called domain walls. Both are expected to be extremely stable and extremely massive. (The monopole can be shown to be about 10^{16} times as heavy as the proton.) A domain of correlated broken-symmetry phase cannot be much larger than the horizon distance at that time, and so the minimum number of defects created during the transition can be estimated. The result is that there would be so many defects after the transition that their mass would dominate the energy density of the universe and thereby speed up its subsequent evolution. The microwave background radiation would reach its present temperature of three degrees K. only 30,000 years after the big bang instead of 10 billion years, and all the successful predictions of the big-bang model would be lost. Thus any successful union of grand unified theories and the big-bang picture must incorporate some mechanism to drastically suppress the production of magnetic monopoles and domain walls.

The inflationary-universe model appears to provide a satisfactory solution to these problems. Before the model can be described, however, we must first explain a few more of the details of symmetry breaking and phase transitions in grand unified theories.

All modern particle theories, including the grand unified theories, are examples of quantum field theories. The best-known field theory is the one that describes electromagnetism. According to the classical (nonquantum) theory of electromagnetism developed by James Clerk Maxwell in the 1860's, electric and magnetic fields have a well-defined value at every point in space, and their variation with time is described by a definite set of equations. Maxwell's theory was modified early in the 20th century in order to achieve consistency with the quantum theory. In the classical theory it is possible to increase the energy of an electromagnetic field by an amount, but in the quantum theory the increases in energy can come only in discrete lumps, the quanta, which in this case are called photons. The photons have both wavelike and particlelike prop-

erties, but in the lexicon of modern physics they are usually called particles. In general the formulation of a quantum field theory begins with a classical theory of fields, and it becomes a theory of particles when the rules of the quantum theory are applied.

As we have already mentioned, an essential ingredient of grand unified theories is the phenomenon of spontaneous symmetry breaking. The detailed mechanism of spontaneous symmetry breaking in grand unified theories is simpler in many ways than the analogous mechanism in crystals. In a grand unified theory spontaneous symmetry breaking is accomplished by including in the formulation of the theory a special set of fields known as Higgs fields (after Peter W. Higgs of the University of Edinburgh). The symmetry is unbroken when all the Higgs fields have a value of zero, but it is spontaneously broken whenever at least one of the Higgs fields acquires a nonzero value. Furthermore, it is possible to formulate the theory in such a way that a Higgs field has a nonzero value in the state of lowest energy density, which in this context is known as the true vacuum. At temperatures greater than about 10^{27} degrees thermal fluctuations drive the equilibrium value of the Higgs field to zero, resulting in a transition to the symmetric phase.

We have now assembled enough background information to describe the inflationary model of the universe, beginning with the form in which it was first proposed by one of us (Guth) in 1980. Any cosmological model must begin with some assumption about the initial conditions, but for the inflationary model the initial conditions can be rather arbitrary. One must assume, however, that the early universe included at least some regions of gas that were hot compared with the critical temperature of the phase transition and that were also expanding. In such a hot region the Higgs field would have a value of zero. As the expansion caused the temperature to fall it would become thermodynamically favorable for the Higgs field to acquire a nonzero value, bringing the system to its broken-symmetry phase.

For some values of the unknown parameters of the grand unified theories this phase transition would occur very slowly compared with the cooling rate. As a result the system could cool to well below 10^{27} degrees with the value of the Higgs field remaining at zero. This phenomenon, known as supercooling, is quite common in condensed-matter physics; water, for example, can be supercooled to more than 20 degrees below its freezing point, and glasses are formed by rapidly supercooling a liquid to a temperature well below its freezing point.

As the region of gas continued to supercool, it would approach a peculiar state of matter known as a false vacuum. This state of matter has never been observed, but it has properties that are unambiguously predicted by quantum field theory. The temperature, and hence the thermal component of the energy density, would rapidly decrease and the energy density of the state would be concentrated entirely in the Higgs field. A zero value for the Higgs field implies a large energy density for the false vacuum. In the classical form of the theory such a state would be absolutely stable, even though it would not be the state of lowest energy density. States with a lower energy density would be separated from the false vacuum by an intervening energy barrier, and there would be no energy available to take the Higgs field over the barrier (see Figure 83, top).

In the quantum version of the model the false vacuum is not absolutely stable. Under the rules of the quantum theory all the fields would be continually fluctuating. As was first described by Sidney R. Coleman of Harvard, a quantum fluctuation would occasionally cause the Higgs field in a small region of space to "tunnel" through the energy barrier, nucleating a "bubble" of the broken-symmetry phase. The bubble would then start to grow at a speed that would rapidly approach the speed of light, converting the false vacuum into the broken-symmetry phase. The rate at which bubbles form depends sensitively on the unknown parameters of the grand unified theory; in the inflationary model it is assumed that the rate would be extremely low. These features are illustrated in Figure 83. There the energy density of the universe is represented by three-dimensional diagrams as a function of two Higgs fields. Each surface shown in cross section is rotationally symmetric about a vertical axis, which corresponds to a state in which both Higgs fields have a value of zero. In the absence of thermal excitations this state of unbroken symmetry, the false vacuum, would have an energy density of about 10^{95} ergs per cubic centimeter, or some 10^{59} times the energy density of an atomic nucleus. The rotational symmetry is broken whenever one of the Higgs fields acquires a nonzero value (or both of them do). Here the theory has been formulated in such a way that the states of lowest energy density,

known as the true-vacuum states, are states of broken symmetry, forming a circle in the horizontal plane at the bottom of each diagram. In this analogy the evolution of the universe can be traced by imagining a ball rolling on the surface. The ball's distance from the central axis represents the combined values of the Higgs fields and its height above the horizontal surface represents the energy density of the universe. When the Higgs fields both have a value of zero, the ball is poised at the axis of symmetry; when the Higgs fields have a value that corresponds to the lowest possible energy density, the ball is lying somewhere in the trough that defines the broken-symmetry, or true-vacuum, states. In the original form of the inflationary-universe model it was assumed that the energy-density function had the shape in the diagram at the top. The inflationary episode would then take place while the universe was in the false-vacuum state. If the laws of classical physics applied, this state would be absolutely stable, because there would be no energy to carry the Higgs fields over the intervening energy barrier. As noted before, the fields in small regions of space can "tunnel" through the energy barrier, forming bubbles of the broken-symmetry phase, which would then start to grow.

The most peculiar property of the false vacuum is probably its pressure, which is both large and negative. To understand why, consider again the process by which a bubble of true vacuum would grow into a region of false vacuum. The growth is favored energetically because the true vacuum has a lower energy density than the false vacuum. The growth also indicates, however, that the pressure of the true vacuum must be higher than the pressure of the false vacuum, forcing the bubble wall to grow outward. Because the pressure of the true vacuum is zero, the pressure of the false vacuum must be negative. A more detailed argument shows that the pressure of the false vacuum is equal to the negative value of its energy density (when the two quantities are measured in the same units).

The negative pressure would not result in mechanical forces within the false vacuum, because mechanical forces arise only from differences in pressure. Nevertheless, there would be gravitational effects. Under ordinary circumstances the expansion of the region of gas would be slowed by the mutual gravitational attraction of the matter within it. In Newtonian physics this attraction is proportional to the mass density, which in relativistic

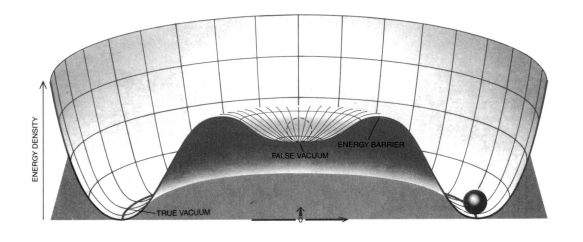

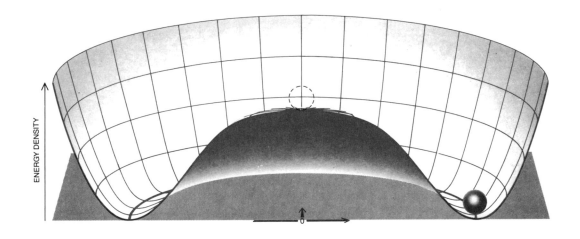

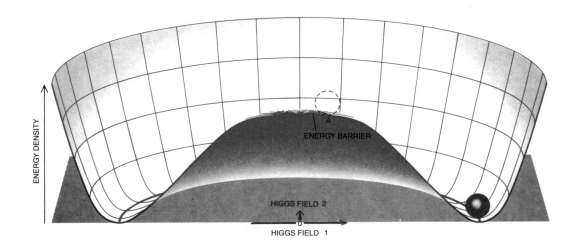

theories is equal to the energy density divided by the square of the speed of light. According to general relativity, the pressure also contributes to the attraction; to be specific, the gravitational force is proportional to the energy density plus three times the pressure. For the false vacuum the contribution made by the pressure would overwhelm the energy-density contribution and would have the opposite sign. Hence the bizarre notion of negative pressure leads to the even more bizarre effect of a gravitational force that is effectively repulsive. As a result the expansion of the region would be accelerated and the region would grow exponentially, doubling in diameter during each interval of about 10^{-34} second.

This period of accelerated expansion is called the inflationary era, and it is the key element of the inflationary model of the universe. According to the model, the inflationary era continued for 10^{-32} second or longer, and during this period the diameter of the universe increased by a factor of 10^{50} or more. It is assumed that after this colossal expansion the transition to the broken-symmetry phase finally took place. The energy density of the false vacuum was then released, resulting in a tremendous amount of particle production. The region was reheated to a temperature of almost 10^{27} degrees. (In the language of thermodynamics the energy released is called the latent heat; it is analogous to the energy released when water freezes.) From this point on the region would continue to expand and cool at the rate described by the standard big-bang model. A volume the size of the observable universe would lie well within such a region.

Both the standard model and the inflationary universe are illustrated in Figure 84. The gray curves represent the standard big-bang model, which is coincident with the inflationary model for all times after 10^{-30} second. For comparison the graph of temperature (top) also includes an indication of the boiling point of water (373 degrees Kelvin) and the temperature at the center of a typical star (10 million degrees K.). Similarly the graph of energy density (middle) indicates the energy density of water (10^{21} ergs per cubic centimeter) and of an atomic nucleus

(10^{36} ergs per cubic centimeter). On the graph of spatial dimensions (bottom) each cosmological model is represented by two curves. One curve shows the region of space that evolves to become the observed universe and the other shows the horizon distance: the total distance a light signal could have traveled since the beginning of the universe. On the time axis several significant events are marked. The line labeled A indicates the time of the phase transition predicted in the standard big-bang model by grand unified theories of the interactions of elementary particles; at the high temperature prevailing before this time the various nongravitational forces acting between particles are thought to have been related to one another by a symmetry that was spontaneously broken when the temperature fell to a critical value of about 10^{27} degrees. A key feature of the inflationary model in the prolongation of the phase transition, which extends through the inflationary era (color band); during the era the universe expands by an extraordinary factor, perhaps 10^{50} or more. Meanwhile the temperature plunges, but it is stabilized at about 10^{22} degrees by quantum effects that arise in the context of general relativity. The gray band labeled B indicates the period when the lightest atomic nuclei were formed, and C indicates the time when the universe became transparent to electromagnetic radiation.

The horizon problem is avoided in a straightforward way. In the inflationary model the observed universe evolves from a region that is much smaller in diameter (by a factor of 10^{50} or more) than the corresponding region in the standard model. Before inflation begins the region is much smaller than the horizon distance, and it has time to homogenize and reach thermal equilibrium. This small homogeneous region is then inflated to become large enough to encompass the observed universe. Thus the sources of the microwave background radiation arriving today from all directions in the sky were once in close contact; they had time to reach a common temperature before the inflationary era began.

The flatness problem is also evaded in a simple and natural way. The equations describing the evolution of the universe during the inflationary era are different from those for the standard model, and it turns out that the ratio Ω is driven rapidly toward 1, no matter what value it had before inflation. This behavior is most easily understood by recalling that a value of $\Omega = 1$ corresponds to a space that is geometrically flat. The rapid expansion causes the space to become flatter just as the surface of a bal-

Figure 83 ENERGY DENSITY OF THE UNIVERSE, shown in three three-dimensional diagrams as a function of two Higgs fields: original form of the inflationary-universe model (top); the new inflationary model (middle); and a variant of the new inflationary model (bottom).

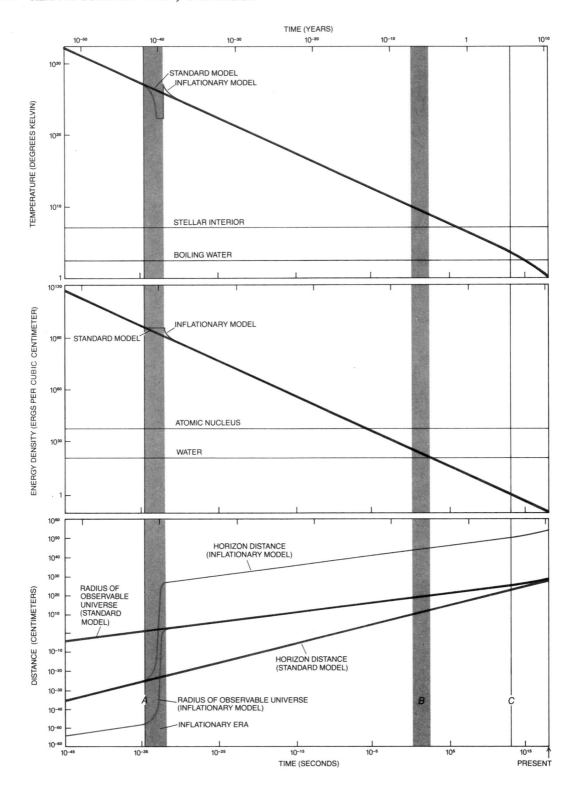

Figure 84 INFLATIONARY MODEL of the universe is represented by the colored curves in this set of graphs, showing how several properties of the observed universe could have changed with time starting at 10^{-45} second after the big bang.

loon becomes flatter when it is inflated. The mechanism driving Ω toward 1 is so effective that one is led to an almost rigorous prediction: The value of Ω today should be very accurately equal to 1. Many astronomers (although not all) think a value of 1 is consistent with current observations, but a more reliable determination of Ω would provide a crucial test of the inflationary model.

The whole picture is somewhat more complicated. Figure 86 illustrates the situation for both the big bang and the inflationary universe. Expanding bubbles of broken-symmetry phase form in an expanding region of symmetric phase in the two highly schematic time sequences. The sequence representing the standard big-bang model (*left*) covers a much shorter time span than the sequence representing the original form of the inflationary model (*right*). In both cases the Higgs fields have a value of zero in the region outside the bubbles, whereas at least one Higgs field has a nonzero value inside each bubble. In a grand unified theory the broken-symmetry states can in general be distinguished by

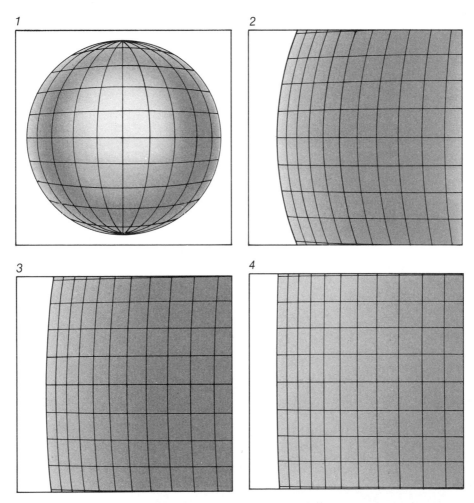

Figure 85 SOLUTION OF THE FLATNESS PROBLEM is illustrated by these perspective drawings of an inflating sphere. In each successive frame the sphere is inflated by a factor of three (while the number of grid lines on the surface is increased by the same factor). The curvature of the surface quickly becomes undetectable on the scale of the illustration.

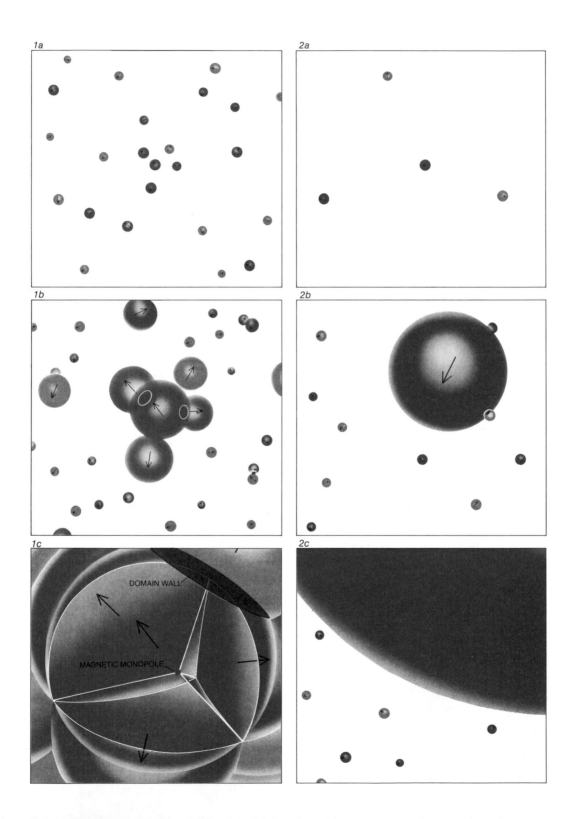

1a

2a

1b

2b

1c

DOMAIN WALL

MAGNETIC MONOPOLE

2c

parameters of two kinds: discrete and continuous. Each bubble is labeled in two ways: by a color (blue or red) to indicate the discrete parameter and by an internal black arrow to indicate the value of the continuous parameter. In the standard model the bubbles would quickly coalesce and complete the transition from the symmetric phase to the broken-symmetry phase. A surfacelike defect, a domain wall, would form at any boundary between regions with different values of the discrete parameter (*purple areas*). Within a region of uniform color a pointlike defect, a magnetic monopole, would form at a center created by the intersection of many bubbles whenever the arrow representing the continuous parameter points everywhere away from the center. In the original form of the inflationary model the rapid expansion of the false-vacuum, or symmetric-phase, region would keep the bubbles from ever coalescing.

In the form in which the inflationary model was originally proposed it had a crucial flaw: under the circumstances described, the phase transition itself would create inhomogeneities much more extreme than those observed today. As we have already described, the phase transition would take place by the random nucleation of bubbles of the new phase. It can be shown that the bubbles would always remain in finite clusters disconnected from one another, and that each cluster would be dominated by a single largest bubble. Almost all the energy in the cluster would be initially concentrated in the surface of the largest bubble, and there is no apparent mechanism to redistribute energy uniformly. Such a configuration bears no resemblance to the observed universe.

For almost two years after the invention of the inflationary-universe model it remained a tantalizing but clearly imperfect solution to a number of important cosmological problems. Near the end of 1981, however, a new approach was developed by A. D. Linde of the P. N. Lebedev Physical Institute in Moscow and independently by Andreas Albrecht and one of us (Steinhardt) of the University of Pennsylvania. This approach, known as the new

inflationary universe, avoids all the problems of the original model while maintaining all its successes.

The key to the new approach is to consider a special form of the energy-density function that describes the Higgs field (see middle diagram of Figure 83). Quantum field theories with energy-density functions of this type were first studied by Coleman, working in collaboration with Erick J. Weinberg of Columbia University. In contrast to the more typical case shown in the top diagram of Figure 83, there is no energy barrier separating the false vacuum from the true vacuum; instead the false vacuum lies at the top of a rather flat plateau. In the context of grand unified theories such an energy-density function is achieved by a special choice of parameters. As we shall explain below, this energy-density function leads to a special type of phase transition that is sometimes called a slow-rollover transition.

The scenario begins just as it does in the original inflationary model. Again one must assume the early universe had regions that were hotter than about 10^{27} degrees and were also expanding. In these regions thermal fluctuations would drive the equilibrium value of the Higgs fields to zero and the symmetry would be unbroken. As the temperature fell it would become thermodynamically favorable for the system to undergo a phase transition in which at least one of the Higgs fields acquired a nonzero value, resulting in a broken-symmetry phase. As in the previous case, however, the rate of this phase transition would be extremely low compared with the rate of cooling. The system would supercool to a negligible temperature with the Higgs field remaining at zero, and the resulting state would again be considered a false vacuum.

The important difference in the new approach is the way in which the phase transition would take place. Quantum fluctuations or small residual thermal fluctuations would cause the Higgs field to deviate from zero. In the absence of an energy barrier the value of the Higgs field would begin to increase steadily; the rate of increase would be much like that of a ball rolling down a hill of the same shape as the curve of the energy-density function, under the influence of a frictional drag force. Since the energy-density curve is almost flat near the point where the Higgs field vanishes, the early stage of the evolution would be very slow. As long as the Higgs field remained close to zero, the energy density would be almost the same as it is in the false vacuum. As in the original scenario, the region would undergo accelerated expansion, doubling in

Figure 86 EXPANDING BUBBLES of broken-symmetry phase in the standard big-bang model (*left*) and the original inflationary model (*right*). Colors indicate the discrete parameters of the broken-symmetry states and arrows indicate the continuous parameters.

diameter every 10^{-34} second or so. Now, however, the expansion would cease to accelerate when the value of the Higgs field reached the steeper part of the curve. By computing the time required for the Higgs field to evolve, the amount of inflation can be determined. An expansion factor of 10^{50} or more is quite plausible, but the actual factor depends on the details of the particle theory one adopts.

So far the description of the phase transition has been slightly oversimplified. There are actually many different broken-symmetry states, just as there are many possible orientations for the axes of a crystal. There are a number of Higgs fields, and the various broken-symmetry states are distinguished by the combination of Higgs fields that acquire nonzero values. Since the fluctuations that drive the Higgs fields from zero are random, different regions of the primordial universe would be driven toward different broken-symmetry states, each region forming a domain with an initial radius of roughly the horizon distance. At the start of the phase transition the horizon distance would be about 10^{-24} centimeter. Once the domain formed, with the Higgs fields deviating slightly from zero in a definite combination, it would evolve toward one of the stable broken-symmetry states and would inflate by a factor of 10^{50} or more. The size of the domain after inflation would then be greater than 10^{26} centimeters. The entire observable universe, which at that time would be only about 10 centimeters across, would be able to fit deep inside a single domain.

In the course of this enormous inflation any density of particles that might have been present initially would be diluted to virtually zero. The energy content of the region would then consist entirely of the energy stored in the Higgs field. How could this energy be released? Once the Higgs field evolved away from the flat part of the energy-density curve, it would start to oscillate rapidly about the true-vacuum value. Drawing on the relation between particles and fields implied by quantum field theory, this situation can also be described as a state with a high density of Higgs particles. The Higgs particles would be unstable, however: they would rapidly decay to lighter particles, which would interact with one another and possibly undergo subsequent decays. The system would quickly become a hot gas of elementary particles in thermal equilibrium, just as was assumed in the initial conditions for the standard model. The reheating temperature is cal-

culable and is typically a factor of between two and 10 below the critical temperature of the phase transition. From this point on, the scenario coincides with that of the standard big-bang model, and so all the successes of the standard model are retained.

Note that the crucial flaw of the original inflationary model is deftly avoided. Roughly speaking, the isolated bubbles that were discussed in the original model are replaced here by the domains. The domains of the slow-rollover transition would be surrounded by other domains rather than by false vacuum, and they would tend not to be spherical. The term "bubble" is therefore avoided. The key difference is that in the new inflationary model each domain inflates in the course of its formation, producing a vast, essentially homogeneous region within which the observable universe can fit.

Since the reheating temperature is near the critical temperature of the grand-unified-theory phase transition, the matter-antimatter asymmetry could be produced by particle interactions just after the phase transition. The production mechanism is the same as the one predicted by grand unified theories for the standard big-bang model. In contrast to the standard model, however, the inflationary model does not allow the possibility of assuming the observed net baryon number of the universe as an initial condition; the subsequent inflation would dilute any initial baryon-number density to an imperceptible level. Thus the viability of the inflationary model depends crucially on the viability of particle theories, such as the grand unified theories, in which baryon number is not conserved.

One can now grasp the solutions to the cosmological problems discussed above. The horizon and flatness problems are resolved by the same mechanisms as in the original inflationary-universe model. In the new inflationary scenario the problem of monopoles and domain walls can also be solved. Such defects would form along the boundaries separating domains, but the domains would have been inflated to such an enormous size that the defects would lie far beyond any observable distance. (A few defects might be generated by thermal effects after the transition, but they are expected to be negligible in number.)

The new inflationary model is illustrated in Figure 87. In the two series of drawings representing the standard big-bang model (*left*) and the new inflationary model (*right*) the gray sphere corresponds to the region of space that evolved to become the observed universe and the two-headed green arrow represents the horizon distance. (The relative scales shown here are suggestive only; the actual scales differ by factors that are too extreme to depict.) Three evolutionary stages are shown for each scenario: just before the phase transition (*top*), just after the phase transition (*middle*) and today (*bottom*). In the standard model the horizon distance is always smaller than the gray sphere, making the large-scale uniformity of the observed universe puzzling. Since in the standard model a domain of broken-symmetry phase created in the phase transition would have a radius comparable to the horizon distance, many monopoles and domain walls would be present in the observed universe. In the new inflationary model the horizon distance is always much larger than the gray sphere, and so the observed universe is expected to be uniform on a large scale and to have few, if any, monopoles and domain walls. Just before the phase transition the gray sphere in the inflationary model is much smaller than it is in the standard model; during the phase transition the gray sphere in the inflationary model expands by a factor of 10^{50} or more in radius to match the size of the corresponding sphere in the standard model.

Thus with a few simple ideas the improved inflationary model of the universe leads to a successful resolution of several major problems that plague the standard big-bang picture: the horizon, flatness, magnetic-monopole and domain-wall problems. Unfortunately the necessary slow-rollover transition requires the fine tuning of parameters; calculations yield reasonable predictions only if the parameters are assigned values in a narrow range. Most theorists (including both of us) regard such fine tuning as implausible. The consequences of the scenario are so successful, however, that we are encouraged to go on in the hope we may discover realistic versions of grand unified theories in which such a slow-rollover transition occurs without fine tuning.

The successes already discussed offer persuasive evidence in favor of the new inflationary model. Moreover, it was recently discovered that the model may also resolve an additional cosmological problem not even considered at the time the model was developed: the smoothness problem. The generation of density inhomogeneities in the new inflationary universe was addressed in the summer of 1982 at the Nuffield Workshop on the Very Early Universe by a number of theorists, including James M. Bardeen of the University of Washington, Stephen W. Hawking of the University of Cambridge, So-Young Pi of Boston University, Michael S. Turner of the University of Chicago, A. A. Starobinsky of the L. D. Landau Institute of Theoretical Physics in Moscow and the two of us. It was found that the new inflationary model, unlike any previous cosmological model, leads to a definite prediction for the spectrum of inhomogeneities. Basically the process of inflation first smoothes out any primordial inhomogeneities that might have been present in the initial conditions. Then in the course of the phase transition inhomogeneities are generated by the quantum fluctuations of the Higgs field in a way that is completely determined by the underlying physics. The inhomogeneities are created on a very small scale of length, where quantum phenomena are important, and they are then enlarged to an astronomical scale by the process of inflation.

The predicted shape for the spectrum of inhomogeneities is essentially scale-invariant; that is, the magnitude of the inhomogeneities is approximately equal on all length scales of astrophysical significance. This prediction is comparatively insensitive to the details of the underlying grand unified theory. It turns out that a spectrum of precisely this shape was proposed in the early 1970's as a phenomenological model for galaxy formation by Edward R. Harrison of the University of Massachusetts at Amherst and Yakov B. Zel'dovich of the Institute of Physical Problems in Moscow, working independently. The details of galaxy formation are complex and are still not well understood, but many cosmologists think a scale-invariant spectrum of inhomogeneities is precisely what is needed to explain how the present structure of galaxies and galactic clusters evolved (see Chapter 3, "The Large-Scale Structure of the Universe," by Joseph Silk, Alexander S. Szalay and Yakov B. Zel'dovich).

The new inflationary model also predicts the magnitude of the density inhomogeneities, but the prediction is quite sensitive to the details of the underlying particle theory. Unfortunately the magnitude that results from the simplest grand unified theory is far too large to be consistent with the observed uniformity of the cosmic microwave background. This inconsistency represents a problem, but it is not yet known whether the simplest grand unified theory is the correct one. In particular the simplest grand unified theory predicts a lifetime for

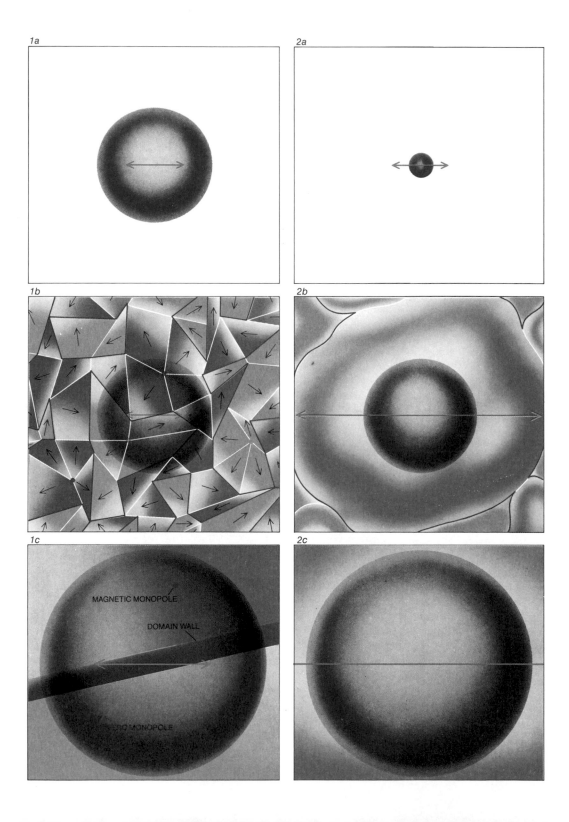

MAGNETIC MONOPOLE

DOMAIN WALL

the proton that appears to be lower than present experimental limits. On the other hand, one can construct more complicated grand unified theories that result in density inhomogeneities of the desired magnitude. Many investigators imagine that with the development of the correct particle theory the new inflationary model will add the resolution of the smoothness problem to its list of successes.

One promising line of research involves a class of quantum field theories with a new kind of symmetry called supersymmetry. Supersymmetry relates the properties of particles with integer angular momentum to those of particles with half-integer angular momentum; it thereby highly constrains the form of the theory. Many theorists think supersymmetry might be necessary to construct a consistent quantum theory of gravity, and to eventually unify gravity with the strong, the weak and the electromagnetic forces. A tantalizing property of models incorporating supersymmetry is that many of them give slow-rollover phase transitions without any fine tuning of parameters. The search is on to find a supersymmetry model that is realistic as far as particle physics is concerned and that also gives rise to inflation and to the correct magnitude for the density inhomogeneities.

In short, the inflationary model of the universe is an economical theory that accounts for many features of the observable universe lacking an explanation in the standard big-bang model. The beauty of the inflationary model is that the evolution of the universe becomes almost independent of the details of the initial conditions, about which little if anything is known. It follows, however, that if the inflationary model is correct, it will be difficult for anyone to ever discover observable consequences of the conditions existing before the inflationary phase transition. Similarly, the vast distance scales created by inflation would make it essentially impossible to observe the structure of the universe as a whole. Nevertheless, one can still discuss these issues, and a number of remarkable scenarios seem possible.

The simplest possibility for the very early universe is that it actually began with a big bang, expanded rather uniformly until it cooled to the critical temperature of the phase transition and then proceeded according to the inflationary scenario. Extrapolating the big-bang model back to zero time brings the universe to a cosmological singularity, a condition of infinite temperature and density in which the known laws of physics do not apply. The instant of creation remains unexplained. A second possibility is that the universe began (again without explanation) in a random, chaotic state. The matter and temperature distributions would be nonuniform, with some parts expanding and other parts contracting. In this scenario certain small regions that were hot and expanding would undergo inflation, evolving into huge regions easily capable of encompassing the observable universe. Outside these regions there would remain chaos, gradually creeping into the regions that had inflated.

Recently there has been some serious speculation that the actual creation of the universe is describable by physical laws. In this view the universe would originate as a quantum fluctuation, starting from absolutely nothing. The idea was first proposed by Edward P. Tryon of Hunter College of the City University of New York in 1973, and it was put forward again in the context of the inflationary model by Alexander Vilenkin of Tufts University in 1982. In this context "nothing" might refer to empty space, but Vilenkin uses it to describe a state devoid of space, time and matter. Quantum fluctuations of the structure of space-time can be discussed only in the context of quantum gravity, and so these ideas must be considered highly speculative until a working theory of quantum gravity is formulated. Nevertheless, it is fascinating to contemplate that physical laws may determine not only the evolution of a given state of the universe but also the initial conditions of the observable universe.

As for the structure of the universe as a whole, the inflationary model allows for several possibilities. (In all cases the observable universe is a very small fraction of the universe as a whole; the edge of our domain is likely to lie 10^{35} or more light-years away.) The first possibility is that the domains meet one another and fill all space. The domains are then separated by domain walls, and in the interior of each wall is the symmetric phase of the grand unified theory. Protons or neutrons passing through such a wall would decay instantly. Domain walls would tend to straighten with time. After 10^{35} years or more smaller domains (possibly even our own) would disappear and larger domains would grow.

Figure 87 NEW INFLATIONARY MODEL. In the standard big-band model (*left*) and the new inflationary model (*right*), the gray sphere corresponds to the region that evolved to become the observed universe. The horizon distance (*green arrows*) in the new inflationary model is always much larger than the sphere, so the observed universe is expected to be uniform on a large scale and have few, if any, monopoles and domain walls.

Alternatively, some versions of grand unified theories do not allow for the formation of sharp domain walls. In these theories it is possible for different broken-symmetry states in two neighboring domains to merge smoothly into each other. At the interface of two domains one would find discontinuities in the density and velocity of matter, and one would also find an occasional magnetic monopole.

A quite different possibility would result if the energy density of the Higgs fields were described by a curve such as the one in the bottom diagram of Figure 83. As in the other two cases, regions of space would supercool into the false-vacuum state and undergo accelerated expansion. As in the original inflationary model, the false-vacuum state would decay by the mechanism of random bubble formation: quantum fluctuations would cause at least one of the Higgs fields in a small region of space to tunnel through the energy barrier, to the value marked A in the illustration. In contrast to the original inflationary scenario, the Higgs field would then evolve very slowly (because of the flatness of the curve near A) to its true-vacuum value. The accelerated expansion would continue, and the single bubble would become large enough to encompass the observed universe. If the rate of bubble formation were low, bubble collisions would be rare. The fraction of space filled with bubbles would become closer to 1 as the system evolved, but space would be expanding so fast that the volume remaining in the false-vacuum state would increase with time. Bubble universes would continue to form forever, and there would be no way of knowing how much time had elapsed before our bubble was formed. This picture is much like the old steady-state cosmological model on the very large scale, and yet the interior of each bubble would evolve according to the big-bang model, improved by inflation.

From a historical point of view probably the most revolutionary aspect of the inflationary model is the notion that all the matter and energy in the observable universe may have emerged from almost nothing. This claim stands in marked contrast to centuries of scientific tradition in which it was believed that something cannot come from nothing. The tradition, dating back at least as far as the Greek philosopher Parmenides in the fifth century B.C., has manifested itself in modern times in the formulation of a number of conservation laws, which state that certain physical quantities cannot be changed by any physical process. A decade or so ago the list of quantities thought to be conserved included energy, linear momentum, angular momentum, electric charge and baryon number.

Since the observed universe apparently has a huge baryon number and a huge energy, the idea of creation from nothing has seemed totally untenable to all but a few theorists. (The other conservation laws mentioned above present no such problems: the total electric charge and the angular momentum of the observed universe have values consistent with zero, whereas the total linear momentum depends on the velocity of the observer and so cannot be defined in absolute terms.) With the advent of grand unified theories, however, it now appears quite plausible that baryon number is not conserved. Hence only the conservation of energy needs further consideration.

The total energy of any system can be divided into a gravitational part and a nongravitational part. The gravitational part (that is, the energy of the gravitational field itself) is negligible under laboratory conditions, but cosmologically it can be quite important. The nongravitational part is not by itself conserved; in the standard big-bang model it decreases drastically as the early universe expands, and the rate of energy loss is proportional to the pressure of the hot gas. During the era of inflation, on the other hand, the region of interest is filled with a false vacuum that has a large negative pressure. In this case the nongravitational energy increases drastically. Essentially all the nongravitational energy of the universe is created as the false vacuum undergoes its accelerated expansion. This energy is released when the phase transition takes place, and it eventually evolves to become stars, planets, human beings and so forth. Accordingly, the inflationary model offers what is apparently the first plausible scientific explanation for the creation of essentially all the matter and energy in the observable universe.

Under these circumstances the gravitational part of the energy is somewhat ill-defined, but crudely speaking one can say that the gravitational energy is negative, and that it precisely cancels the nongravitational energy. The total energy is then zero and is consistent with the evolution of the universe from nothing.

If grand unified theories are correct in their prediction that baryon number is not conserved, there is no known conservation law that prevents the observed universe from evolving out of nothing. The inflationary model of the universe provides a possible mechanism by which the observed universe could have evolved from an infinitesimal region. It is then tempting to go one step further and speculate that the entire universe evolved from literally nothing.

POSTSCRIPT

Our article described the inflationary universe, a model in which physical processes are used to explain not only the matter-antimatter imbalance, but also the production and primordial distribution of essentially all matter and energy in the universe. As we discussed, this does not violate the conservation of energy—the cosmic gravitational field provides a negative contribution to the total energy, which maintains a constant value at or near zero. The model explains a number of previously mysterious features of the universe and makes several predictions that are in principle testable.

Since our article there have been a number of relevant developments in both particle physics and cosmology. In particle physics a large fraction of the theoretical effort has shifted to the study of "superstrings," an entirely new type of theory that can perhaps provide a unified description of all the known interactions of physics, including gravity. While these theories lie outside the framework of the quantum field theories we discuss, at temperatures small compared to 10^{32} degrees K. they are expected to closely approximate the behavior of supersymmetric grand unified field theories. Thus the process of inflation, believed to begin at about 10^{27} degrees K., could take place as described. In principle the superstring theory will determine the free parameters of the approximate grand unified theory, but in practice these low-energy consequences of superstring theories are difficult to extract. No one yet knows whether the theories are compatible in detail with either particle-physics phenomenology or with inflationary cosmology. Nonetheless the theories are very promising.

Work has continued on the construction of detailed grand unified theories that might underlie inflationary cosmology. For the predicted density inhomogeneities to be acceptably small, the field that drives the inflation must interact weakly with the other fields and with itself. For this reason it now appears that this field cannot be the strongly interacting Higgs field required for the breaking of the symmetry between the strong, weak and electromagnetic interactions. It must instead be a new field, with an energy density diagram of the form we discussed. This extra field is not a serious problem for the model, since many recent particle and superstring theories predict the existence of weakly interacting fields that may have the properties needed for driving inflation. However, the extreme weakness of the interactions implies that such a field would not have had time to reach a state of thermal equilibrium before the temperature fell to 10^{27} degrees K. Nonthermal initial states have been shown to produce inflation from initial situations chosen randomly according to certain specified procedures. Motivated in part by this failure to attain thermal equilibrium, Linde and his collaborators have pursued the idea of "chaotic inflation," in which both the inflation-driving field and the geometry of space are assumed to begin in a totally chaotic state. This approach has the advantage that there is a larger class of particle theories for which the scenario is workable. No one, however, has quantified the range of initial configurations that lead to inflation, so it is difficult to evaluate the plausibility of the inflationary model under such random initial conditions.

Work has also continued on the distribution of matter in the universe, but our understanding remains incomplete. In particular, it is still not clear whether the mean mass density of the universe is equal to the critical value, as predicted by inflationary cosmology. The mass density of visible stars is only a few percent of the critical density, but measurements of the motion of stars within galaxies and galaxies within clusters indicate that the gravitational binding of these structures requires a much larger amount of "dark matter," discussed in Chapter 1, "Dark Matter in the Universe," by Lawrence Krauss. If one assumes that the dark matter is distributed in the same pattern as the visible galaxies, then one infers a mean mass density of a few tenths of the critical value. Some argue, however, that the dark-matter distribution is very different from that of visible galaxies, with galaxies forming only in regions of unusually large mass density. In models of this type a critical mean mass density is consistent with observations. In addition, several new methods for measuring the mean mass density suggest a value near the critical density.

The nature of the dark matter remains a mystery. Big-bang nucleosynthesis calculations imply that the total mass in baryonic matter (that is, protons and neutrons) is no more than one-seventh the critical density. Therefore critical density would require large amounts of nonbaryonic dark matter, most likely in the form of some weakly interacting particle. Massive neutrinos are one possibility, but a variety of hypothetical particles suggested by recent particle theories seem to better fit the phenomenology of galactic structure. A more radical proposal is that the vacuum has a nonzero mass density, producing a total density equal to the critical value. In general relativity such a vacuum mass density is equivalent to a positive value of Einstein's cosmological constant, which is normally assumed to be zero.

The form of the spectrum of initial density inhomogeneities has also been a subject of active study, but it is still not known whether the scale-invariant spectrum predicted by the inflationary model is compatible with observation. Recently there has been promising work on another mechanism for generating inhomogeneities, based on linelike defects in the pattern of spontaneous symmetry breaking called "cosmic strings." While inflation is still required in these models, the parameters are chosen so that the density inhomogeneities created during inflation are negligibly small. The dominant inhomogeneities are generated by the random formation of cosmic strings in a phase transition that takes place after the inflationary era.

The general features of the inflationary universe model are highly attractive and have become a working hypothesis in cosmological research. The details of the model, however, are intimately linked to uncertainties in particle physics and astrophysics and are likely to present challenges for years to come.

The Future of the Universe

A forecast for the expanding universe through the year 10^{100}: All protons will decay, galaxies will form black holes and black holes will "evaporate." If the universe collapses, it may cycle.

. . .

Duane A. Dicus, John R. Letaw, Doris C. Teplitz and Vigdor L. Teplitz

In the past dozen years developments in the physics of high-energy interactions of elementary particles have stimulated a remarkable growth in cosmology. The attempt to describe all the fundamental forces of nature as different aspects of one underlying force has had partial success in the "grand unified theories" of elementary particles. Through such theories it is possible to give tentative descriptions of dominant physical processes from extremely low temperatures to temperatures on the order of 10^{32} degrees Kelvin. The properties of matter at densities of cosmological interest, which vary from less than 10^{-300} gram to more than 10^{100} grams per cubic centimeter, can thereby be addressed. The extreme conditions at the ends of these scales can prevail only in the very early or the very late stages in the evolution of the universe.

A decade ago cosmologists began to show how theoretical physics can give a consistent account of the history of the universe to within 10^{-35} second of its beginning. More recently several physicists and cosmologists, including the four of us, have extrapolated cosmic events into a future that is up to 10^{100} times the current age of the universe. Apart from the intrinsic fascination of such an enterprise, there are several reasons for undertaking it. From the standpoint of theoretical physics the extrapolation makes possible a kind of thought experiment: By determining how the diverse effects predicted by a theory interact at a given time, one can test the overall consistency and plausibility of the theory. No ordinary laboratory is ever likely to attain the temperatures and densities at which grand unified theories make predictions, and so the theories are being verified in what some physicists call the ultimate laboratory, namely the universe as a whole. From the standpoint of cosmology the importance of our extrapolations is that the grand unified theories also have consequences that can be tested in terrestrial laboratories, and so the predictions they make under extreme conditions can be confirmed. One can therefore delineate a sequence of events in the distant future with considerably more detail than has previously been possible.

The framework for our calculations about the remote future is the big-bang model. There is now general agreement that the evolution of the universe must have been determined in the first few moments of the big bang. According to the model, the universe began as the explosion of a highly compact, primordial entity some 10 to 20 billion years ago. The term big bang and the terminology associated with an explosion are appropriate in part, because matter and energy do indeed appear to have been flung outward in space from a common origin. The term can also be misleading, however, because it is not correct to suppose the expansion can be visualized from the outside. There is no external vantage because what is exploding is the entire universe. Space itself is expanding in the sense that the separation between any two galaxies is

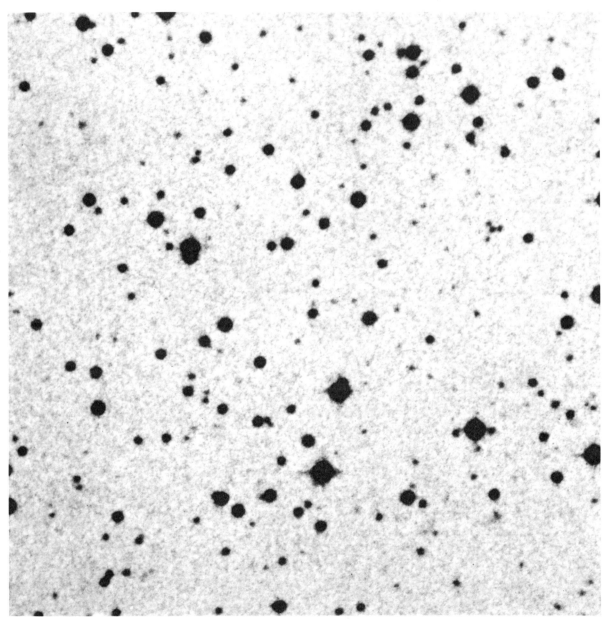

Figure 88 MOST DISTANT OBJECT KNOWN is a quasar (the dark, starlike object in center of this negative print) designated PKS 2000-330; its red shift of 3.78 implies it is receding from us at 92 percent of the speed of light. It is estimated to be more than 16 billion light-years from our galaxy. Light now reaching the earth from the quasar must have been emitted shortly after the big bang.

growing at a rate that depends on their separation. Viewed from our galaxy all other galaxies appear to be receding at a rate that increases by 17 kilometers per second for every million light-years' separation. Mathematically the big bang is described by Einstein's general theory of relativity.

The recession of distant galaxies is inferred from the observed shift of their radiation toward the red end of the electromagnetic spectrum. The expansion and cooling of the universe has altered even more dramatically the spectrum of another kind of radiation, namely the cosmic microwave background radiation, discovered in 1964 by Arno A. Penzias and Robert W. Wilson of Bell Laboratories. The background radiation seems to be coming with equal intensity from every point in the sky; the intensity varies with direction by less than one part in 10,000. It was the scattering and annihilation of electrically charged particles in the early stages of the big bang that gave rise to the background radiation. The radiation was therefore once very hot, but it has since been cooled by the expansion of the universe to a temperature of about three degrees K.

Nuclear reactions at the density and temperature of the universe about three minutes after the start of the big bang were responsible for the synthesis of helium and, to a much lesser extent, other light elements. The universe cooled too quickly, however, for carbon and heavier elements to form, and a generous supply of hydrogen was left over to serve as nuclear fuel for stars. Earlier, reactions only 10^{-38} second after the universe began may have generated the observed excess of matter over antimatter, in which case most grand unified theories predict the eventual decay of all nuclear matter. The current understanding of the conditions in the early universe has not, however, been sufficient to answer the most compelling question in cosmology: Will the universe continue to expand forever, or will gravitational forces halt the expansion and drag all space and time back into the fireball with which the universe began?

Although many years of observational effort and theoretical consideration have been given to the last question, the issue is still undecided. In principle the simplest way to resolve it is to measure the amount of matter in a large, representative volume of space and so estimate the density of matter in the universe at large. If the density is equal to or greater than a critical density, which is about three protons for every 1,000 liters of space, gravity will eventually overcome the expansion and the uni-

verse will collapse. A universe of this kind is said to be closed. If the density is less than the critical value, the universe will remain open; the expansion will be slowed by gravity but never stopped. Note that physical processes such as stellar burning or the decay of elementary particles cannot change a closed universe into an open one.

Estimates of the number of stars in a typical galaxy and the number of galaxies per unit volume give a value for the density of matter that is only 5 percent of the critical density. On this basis one might conclude that the universe is open and will continue its expansion forever. An estimate of mass based on the counting of stars, however, includes only luminous matter; if nonluminous matter accounts for a significant fraction of the mass of the universe, the estimate is not accurate.

There is some evidence of nonluminous matter in galaxies. The speed with which a star revolves about the center of its galaxy should increase with distance from the galactic center for stars lying in the interior of the galaxy and should decrease with distance if the star is at the periphery. For outlying stars, however, the expected decrease in speed of revolution is not observed; the observed distribution of speeds leads to the conclusion that at the periphery of a galaxy there is a halo of nonluminous matter whose total mass is one or two times that of the luminous galactic matter (see Figure 89). Hence nonluminous matter within galaxies may account for from 5 to 10 percent of the mass needed to close the universe.

Similarly, the observed motions of galaxies in clusters of galaxies appear to require enough nonluminous matter to make up 50 percent of the critical density of the universe. The nonluminous matter would explain why the galaxies remain bound in clusters instead of moving off in random directions. The dark matter could be in the form of cold rocks, burned-out stars, clouds of gas or even black holes.

If the additional matter necessary to make up the remaining 50 percent of the critical density actually exists, it may consist of massive neutrinos or of any of a number of exotic elementary particles that have been postulated by recent theories. For neutrinos there should be about as many of them as there are photons, or quanta of electromagnetic energy, associated with the cosmic background radiation; this amounts to a few hundred neutrinos per cubic centimeter. Their contribution to the density of the universe depends on the mass of the neutrino. It was once supposed the neutrino is massless, but current experiments only place upper bounds on neutrino

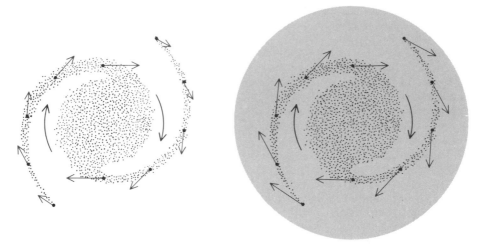

Figure 89 NONLUMINOUS MATTER may make up half of the mass of most galaxies. If galaxies were made up only of visible matter, stars at a galaxy's apparent periphery would revolve about the galactic center with a velocity that decreases with distance from the center (*left*). However, such a decrease is not observed. If galaxies had mas-sive nonluminous halos, stars that appear to be on the periphery would actually be in the midst of galactic mat-ter, and the distribution of their velocities would more closely match the observed distribution (*right*). The lengths of the arrows are proportional to the velocities of the stars.

masses and do not rule out the possibility that neu-trino masses are sufficient to close the universe. Moreover, certain recent theoretical arguments sug-gest that the neutrino has some mass, and several groups of experimenters are trying to detect and measure it. Such experiments could definitely an-swer the question of closure in the affirmative within the next few years. A negative result, on the other hand, would leave the question to be resolved by more precise measurement of the distances and recessional velocities of remote galaxies.

Although the current uncertainty about the den-sity of the universe straddles the value that distinguishes a closed universe from an open one, the uncertainty is confined to less than two orders of magnitude. The fact that the universe has not contracted before reaching its present age implies that the density is not greater than about 10 times the critical density. The estimate of density based on luminous matter alone indicates that the actual den-sity must be at least a twentieth of the critical den-sity. No one yet knows how to account for what cosmologists find to be a striking coincidence.

Since the question of closure has not yet been settled on experimental or observational grounds, both the open and the closed possibilities need to be considered in any speculations about the distant future. Suppose, to begin, that the critical density is

not attained and the universe is open. What is likely to happen? The question must be answered both for large-scale structure (that is, for the evolution of the geometric properties of the universe) and for local composition (for aggregates of matter that range in size from protons to galaxies).

According to our best understanding, the local composition of the open universe will undergo six major transitions. First, within 10^{14} years after the big bang all stars will have run out of fuel. The principal nuclear fuel is hydrogen, which is con-verted into helium in the core of a star throughout most of its lifetime. After most of the hydrogen fuel has been consumed the star rapidly swells to many times its former size and becomes a red giant; in this stage of the star's existence helium is usually con-verted into carbon and heavier elements. These thermonuclear reactions are effectively one-way transitions: hydrogen is converted into helium, he-lium into carbon and carbon into heavier elements in a series that generally terminates with iron. The iron nucleus has a lower total energy per unit of mass than any other nucleus, so that once the "iron limit" has been reached the energy of the universe that is stored as nuclear fuel will have been entirely released.

The rate at which nuclear fuel is consumed by a star depends on the mass of the star: the more massive the star is, the faster it burns and so the shorter its lifetime is. The sun, for example, will use up much of its hydrogen within about 10 billion years, whereupon it will rapidly fluctuate in size, burn through some of the heavier elements at a prodigious rate but for a relatively short time and finally collapse into a small, slowly cooling white-dwarf star. The smallest stars will require many times the life span of the sun to pass through all these stages, but eventually they too will reach the iron limit. Note that although the exhaustion of nuclear fuel is the first major event in the future of the open universe, it will happen at a time exceedingly remote from our own. The last stars will cease to shine only after the universe is 10,000 times as old as it is now.

The second major event is that all stars will lose their planets. If a star that has a planet is approached by another star to within the radius of the planetary orbit, the orbit will be markedly disrupted by the gravitational field of the passing star, and the planet can be scattered into space. The average time before such an encounter can be expected depends on the density of stars in a given region, on the area of the planetary orbits and on the velocity of the stars with respect to one another. The density of stars can be expressed in terms of the volume in which at least one star is likely to be found. A star moving through space with a planet traces a cylindrical volume whose size depends on the area of the orbit of the planet and on the star's speed. The average interval between stellar encounters is equal to the time required for the volume of the cylinder to become as large as the volume in which a star can be expected (see Figure 90).

The density of stars in a galaxy is one star per 35 cubic light-years. Freeman J. Dyson of the Institute for Advanced Study in Princeton suggests that a

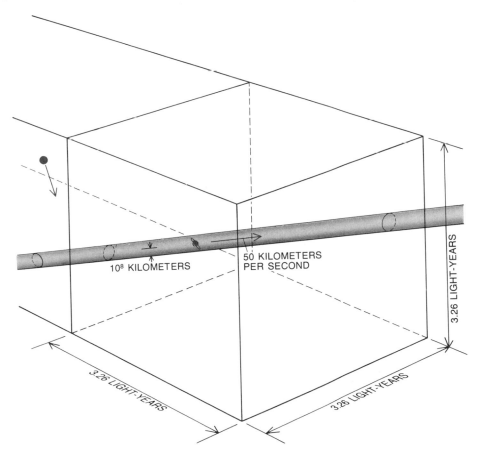

Figure 90 FREQUENCY OF ENCOUNTERS between a star and a planetary system can be estimated from the number of stars per unit volume, the size of the planetary system and its velocity with respect to the stars. The scale of the stars and the planetary system with respect to their density in space has been greatly exaggerated.

reasonable value for the radius of a planetary orbit is about 100 million kilometers, and he notes that a typical star pulls its planets through space at about 50 kilometers per second. The volume of the cylin- der traced by the moving planetary system will therefore be 35 cubic light-years after about 10^{15} years, and so a disruptive encounter is likely in this interval. It can be safely assumed that after 100 such

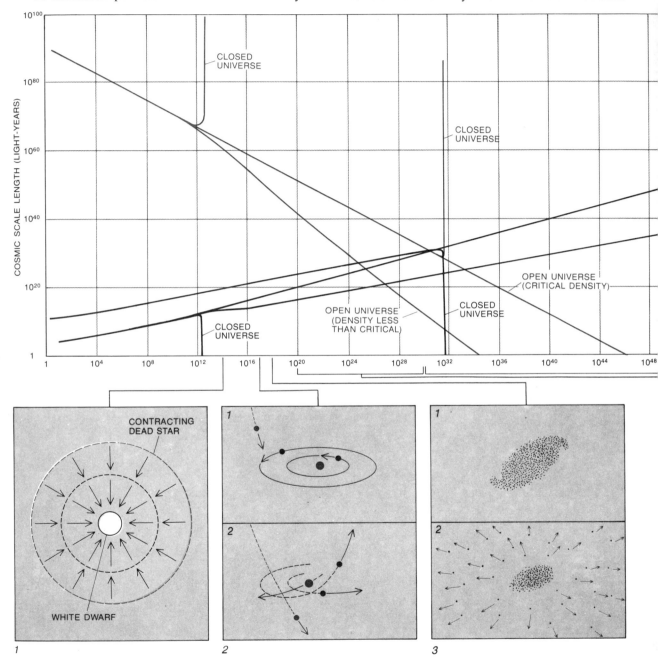

Figure 91 SIX MAJOR EVENTS in the future of the uni- verse (*bottom drawings*) and the density and cosmic scale factor for four models of the universe plotted against time:

(1) stars run out of fuel (10^{14} years); (2) stars lose their planets (10^{17} years); (3) galaxies lose a large fraction of their mass (10^{18} years) and the remaining matter collapses into a

encounters all of a star's planets will have been kicked out of orbit, so that in about 100 times 10^{15} years, or 10^{17} years, all stars will have lost their planets.

The third transition foreseen is the result of still closer stellar encounters; their effects are evident on a galactic scale. When two stars come close to each other, the gravitational interaction can transfer ki-

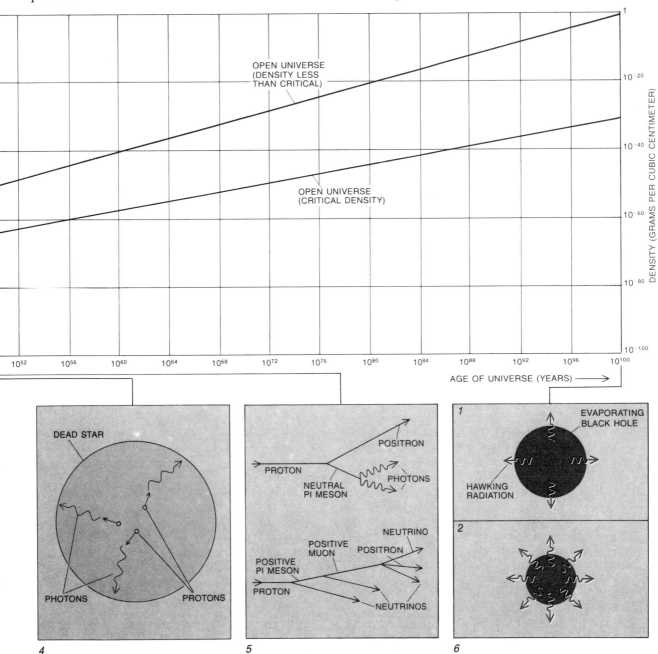

supermassive black hole; (4) heat generated by decay of protons inside the cold stars becomes significant (10^{20}) years; (5) 40 percent of the matter in the universe undergoes such decay (10^{30}) years; and (6) after 10^{100} years black holes lose their mass through quantum evaporation.

netic energy from one star to the other. If the encounter is close enough, one star can gain so much energy that it attains the velocity needed to escape from the galaxy. Because energy is conserved in the interaction, the second star in the pair must lose kinetic energy, and as a result it becomes more tightly bound to the core of the galaxy.

The process can be called galactic evaporation, because the stellar interactions reproduce on a majestic scale the interactions of molecules evaporating at the surface of a liquid. A similar exchange of energy can also cause an appreciable fraction of the interstellar gas to escape from the galaxy. After as much as 90 percent of the mass in the galaxy has evaporated, the gravitational field will draw the remaining stars and dust into an increasingly dense core. Galaxies in their present form are likely to include a central, supermassive black hole, a region of space from which escape is impossible (except through mechanisms described by the quantum theory). If no such black hole exists now, the galactic core would probably still grow so dense that gravitation would overwhelm the resistance offered by gas pressure, and the core would collapse anyway to form a supermassive black hole. A calculation similar to the one we have outlined for the loss of planets shows that the evaporation of stars and the collapse of galaxies should be complete after about 10^{18} years.

The fourth and fifth transitions we envision for the open universe are late cosmological effects predicted by most grand unified models of particle physics, but the effects do not become important until the universe is 100 times older than it is when galaxies collapse. The aim of the grand unified theories is to present a comprehensive account of the strong, the weak and the electromagnetic forces between elementary particles. At the relatively low energies accessible in terrestrial laboratories the three forces seem quite different; nevertheless, the theories describe them as manifestations of a single interaction whose unity can be perceived at the energy corresponding to a temperature on the order of 10^{27} degrees K. At that extraordinary temperature particles such as the quarks, which "feel" the strong force, can be transformed into particles such as the electron and the positron, which at lower energies feel only the weak and the electromagnetic forces. At lower temperatures such changes of identity become highly improbable, but they can still take place occasionally. The proton, which is thought to

be a bound system of three quarks, could decay if its constituent quarks underwent such a transformation.

According to most grand unified theories, all protons should decay after a period of from 10^{30} to 10^{32} years; a decay rate of at least one proton per year should therefore be registered in any aggregate that includes 10^{32} protons, such as a mass of water weighing about 160 tons. In 1983 there were 13 experiments either in operation or under consideration that attempted to monitor large masses of water, iron or other materials for evidence of proton decay. Decay signals had tentatively been identified from 150 tons of iron being monitored 2,300 meters underground at the Kolar goldfields near Bangalore in southern India. A single candidate decay signal had also been registered by detectors installed in the Mont Blanc Tunnel between France and Italy. The postscript addresses whether these signals continued to stand up to scrutiny.

If the proton is susceptible to decay, the process will have a significant effect on stars that have not been captured by galactic black holes. Such stars are the ones that escape by evaporation, and the decay of the protons and neutrons in them will keep them considerably warmer than the surrounding interstellar medium. For an assumed proton lifetime of 10^{30} years the decay rate in a typical star the size of the sun is about 10^{27} per year. Each proton decay gives rise to a shower of energetic electrons, positrons, neutrinos and photons. All the daughter particles except the neutrinos are absorbed by the star, and the absorbed energy keeps the star warm.

The precise temperature of the star during the era of proton decay can be determined by setting the rate at which energy is radiated from the star equal to the rate at which the decay releases heat energy. In this state of equilibrium the temperature depends on the mass of the star, on the surface area from which the heat can be radiated and on the rest energy and lifetime of the proton. The equilibrium temperature is 100 degrees K. for the most massive dead stars (which, paradoxically, are the smallest ones) and about three degrees K. for larger and less massive ones.

The stars will cool to their equilibrium temperature in about 10^{20} years and thereafter their temperature will remain roughly constant until most protons have decayed, in about the year 10^{30}. The stellar glow is cold, but not in comparison with the temperature of the background radiation left over from the big bang. The background temperature depends on the details of the expansion of the open universe. If the density of the universe is less than the critical density, the background-radiation temperature will have fallen to 10^{-20} degree K. by the year 10^{30}. On the other hand, if the density is exactly equal to the critical density, the universe will have expanded more slowly and the temperature of the background radiation will have fallen to 10^{-13} degree K. That is, the temperature will be from 13 to 20 orders of magnitude less than the temperatures of the dead stars.

Proton decay also changes the constitution of interstellar gases that evaporate before galaxies collapse. Inside a star a positron released by the decay of a proton soon encounters an electron, and the two particles annihilate each other. The annihilation generates more photons and heats the star. In intergalactic space the density of matter is so low (and continuously decreasing because of the expansion of the universe) that positrons and electrons are extremely unlikely to collide. Indeed, by the year 10^{30} an open universe with a subcritical density will have expanded to more than 10^{20} times its present size, and the average distance between an electron and a positron in interstellar space will be of the same order of magnitude as the size of our galaxy. (If the density of the universe is equal to the critical density, the universe will have expanded by a factor of 10^{13}.) The interstellar medium therefore becomes an extremely rarefied gas, formed from about 1 percent of today's electrons and from positrons created by the decay of about 1 percent of today's protons.

The events associated with the decay of the proton will have played themselves out by the time the universe has existed for 10^{32} years, or 100 times as long as the average proton lifetime. What remains of the universe then is the rarefied electron-positron gas, the photons and neutrinos that were emitted in various earlier epochs and the supermassive black holes. The photons and neutrinos are left over from the big bang, from the years when the stars shone, from the decay of protons and neutrons throughout history and from the final decay of dead stars. Photons and neutrinos lose energy, and both they and the other constituents become more rarefied as the expansion continues. In other respects the universe at subcritical density remains quiescent until roughly the year 10^{100}, a period lasting 10^{68} times as long as all the processes we have described so far.

The sixth and last major event in the future of the open universe is the decay of the black holes. In the simplest interpretation of Einstein's gravitational theory nothing can escape from a black hole. There is a boundary called the event horizon where the velocity needed to escape is equal to the speed of light; no particle inside the event horizon can ever go fast enough to cross it. Nevertheless, in 1974 S. W. Hawking of the University of Cambridge showed that a quantum-mechanical phenomenon implies that a black hole can give up all the energy associated with its mass and thereby disappear.

The quantum phenomenon is a consequence of the uncertainty principle of Werner Heisenberg, which states that it is only with a certain probability that either the momentum or the position of any particle can be determined to be within a given range of values. More precisely, the product of the uncertainty in the momentum of the particle and the uncertainty in its position is not less than a numerical constant. In classical physics (that is, in physical theory that does not take account of quantum phenomena) a particle can cross an energy barrier only if it is given enough energy to surmount the barrier. The event horizon is an absolute energy barrier in classical physics: a particle could not gain enough energy to surmount it. Because of the uncertainty principle, however, a particle initially found in one region can later be found in another region, even if its classical energy is much less than the height of the energy barrier between the two regions. A particle that crosses an energy barrier in this way, without acquiring the energy necessary to surmount it, is said to have tunneled through the barrier. Hawking pointed out that because particles can tunnel through the event horizon, mass and energy can be ejected from a black hole [see "The Quantum Mechanics of Black Holes," by S. W. Hawking; SCIENTIFIC AMERICAN, January, 1977].

Hawking showed that the rate at which energy is emitted by the black hole is inversely proportional to the square of its mass. The rate is initially low, but as the mass decreases, the energy loss accelerates. It follows that all black holes must eventually disappear, or evaporate; for the supermassive black-hole relics of collapsed galaxies the evaporation time is about 10^{100} years. Most of the decay products are photons; the emissions become increasingly energetic during the later stages of the evaporation. After 10^{100} years, therefore, the universe is made up of an extremely diffuse gas of electrons, positrons and neutrinos, low-energy pho-

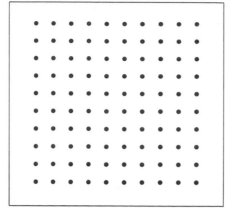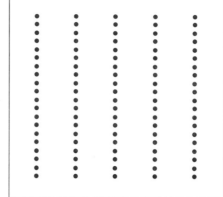

Figure 92 DISTRIBUTION OF MATTER AND ENERGY in the universe appears to be both homogeneous and isotropic. In a homogeneous universe equal numbers of galaxies fill comparable volumes of space; in an isotropic universe the density of matter and energy is the same in all directions as well as being homogeneous. If matter is distributed as in the left diagram, the universe is both homogeneous and isotropic; the distribution on the right is homogeneous but not isotropic.

tons emitted long before black-hole decay and numerous expanding spheres of high-energy photons emitted by evaporating black holes.

Mathematical models of the large-scale structure of an open universe suggest that the universe we live in is a very special place. To describe the universe mathematically six quantities must be specified: three of them give the rate of expansion along the three directions in space and three give the rate of change in the expansion rate (that is, its acceleration or deceleration) along the three directions. If the six quantities are specified at some definite time and if the distribution of energy is given for the same time, the values of the six quantities and the distribution of energy are determined for all later times.

In 1973 Hawking and C. B. Collins, also of Cambridge, showed that for almost all possible values of these quantities at an early stage in the big bang the universe must become increasingly anisotropic as it expands (see Figure 92). The average separation of galaxies along one spatial axis ought to become increasingly different from the average separation along the other axes because of differences in the expansion rates. The result suggests that the actual universe is a highly improbable one. For example, as we noted above, the cosmic microwave background is invariant with direction to within one part in 10,000. On a cosmic scale the matter in the universe is also smoothly distributed. A major unsolved problem in cosmology is to understand why the universe is so nearly isotropic.

One approach to the problem is to choose initial values of the six quantities in such a way that the energy density of matter and radiation is exactly equal to the critical density. If such a universe is dominated by matter instead of radiation throughout its history, any initial anisotropy will not grow. The decay of the proton, however, can lead to a radiation-dominated universe.

John D. Barrow and Frank J. Tipler of the University of California at Berkeley have shown that in a universe at the critical density the electrons and positrons remaining after proton decay will begin to form bound pairs after more than 10^{70} years. Each bound pair is effectively an atom, but the electron and positron orbit each other in such a highly excited state that the region of space encompassed by the pair is larger than the present observable universe. The encompassed space grows progressively larger the later a bound pair is formed.

As time goes on the electron and the positron spiral inward and eventually annihilate each other, giving rise to high-energy photons. The wavelength of the photons is shifted toward the red end of the spectrum as the universe continues to expand. In the model described by Barrow and Tipler the universe finally comes to be dominated by radiation and the anisotropy increases. In a subsequent investigation, however, Don N. Page and M. Randall McKee of Pennsylvania State University showed that under certain conditions the electron-positron pairs can continue to dominate the energy of the universe even though their number is decreasing. The red shift of the photon radiation decreases its energy and the increasingly long time required for the annihilation of late-forming bound pairs causes a decrease in the rate of photon emission. Hence the anisotropy that is characteristic of a radiation-dominated universe may not be able to grow.

All the foregoing speculations concern an open universe. We shall now consider the future of the universe on the assumption that there is enough nonluminous matter for gravity to halt the expansion and bring about a contraction. The closer the average density is to the critical density, the longer the closed universe will last. The universe, if closed, will, if it lives long enough at its maximum expan-

sion, be made up of dead stars, supermassive black-hole remnants of collapsed galaxies and low-energy photons and neutrinos, just as it would be if the universe were open.

One curious aspect of a closed universe is that although energy is conserved locally, the total mass or energy of the universe is not conserved. For a given size of the universe the total energy during contraction is greater than it is during expansion. Consider a photon emitted from the sun toward intergalactic space. The conservation law is satisfied during the emission because the energy carried by the photon is exactly balanced by a very small decrease in the mass of the sun. As the universe expands, the wavelength of the photon increases in proportion, with the result that its energy diminishes. When the universe contracts, the photon's wavelength contracts as well and its energy in-

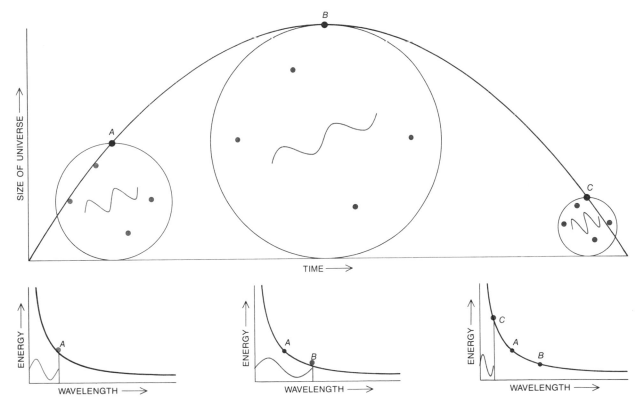

Figure 93 ENERGIES OF FREE PARTICLES are not conserved in a closed universe. Because of cosmic expansion and contraction in such a universe, phenomena that depend on distance vary with time, just as the distance be- tween two points on the surface of a balloon increases or decreases as the balloon is inflated or deflated. Thus the wavelength of a photon increases during expansion and decreases during contraction.

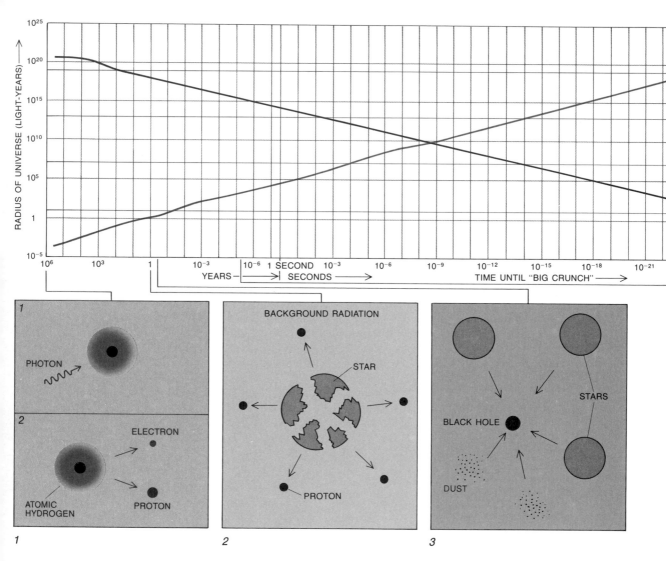

Figure 94 FINAL STAGES in the collapse of a closed universe. About 20 billion years before the big crunch, or complete gravitational collapse, the universe will contract to a volume at which the energy density will be the same as now. Photons will become more energetic and the universe will heat up; about a million years before the crunch the photons will dissociate interstellar hydrogen atoms into their constituent electrons and protons (1). A year before the crunch the temperature outside stars will become greater than inside and the stars will begin to break

creases. Eventually the wavelength of the photon becomes shorter than it was at the time of emission, so that the photon has gained energy without any compensatory loss of mass or energy in some other region. The universe therefore runs hotter in contraction than it does in expansion. The photons that make the largest contribution to the extra energy are the ones emitted when the universe is at or near its maximum size, because such photons will undergo the most contraction.

The major events during the expansion phase of a closed universe follow the same sequence as the events in an open universe. (There is, of course, no collapse for an open universe.) Several investigators have studied the collapse, including Martin J. Rees of Cambridge. As the photons gain energy during the contraction they heat the dead stars, causing them to begin to burn rapidly, explode or evaporate. The resulting soup of particles would continue to retrace the steps in the expansion of the universe if

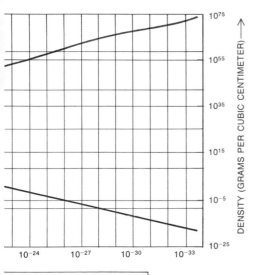

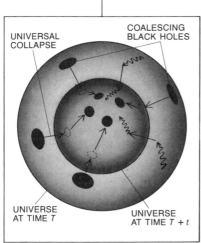

4

up (2). Supermassive black holes will begin to swallow up the remaining stellar material and other matter and radiation (3). The black holes will begin to coalesce about three minutes before the crunch (4). The variation in density and radius of the universe is graphed with time.

it were not for the effects of the black holes. As the density increases, the black holes gobble up material, and they coalesce whenever they collide. One can calculate that in a universe with one supermassive black hole per galaxy the dead stars are swallowed by black holes shortly after the stars break up and material begins to evaporate from them. All the black holes finally coalesce into one large black hole that is coextensive with the universe (see Figure 94).

Theoretical physics in its current state of development cannot fully describe the collapse of a black hole; it is not yet known how to extrapolate the equations that govern the large-scale structure of the universe back to a state of infinite density. It is possible, however, that before the density becomes infinite an unknown mechanism could cause the universe to "bounce" and begin expanding once again.

If the universe can bounce, it could remain closed and so be cyclic. The energy gained by the photons during each period of contraction might be conserved through the bounce; with each successive cycle the universe would therefore be larger, for a given temperature, than it was in the previous cycle and would also take longer to reach its maximum size. Robert H. Dicke and P. J. E. Peebles of Princeton University and the four of us have calculated that, if our universe is cyclic, the next cycle should expand for about twice as long as the expansion phase of the current cycle. For earlier cycles the expansion factor would be smaller; we calculate that the present universe is at most 100 bounces from the cycle that lasted just long enough to make a single generation of stars. The major difficulty for the bounce model is to understand how an extraordinarily inhomogeneous and locally anisotropic universe of isolated, coalescing black holes can be smoothed out; for that matter, there is no understanding of how a bounce can take place.

If the universe is cyclic, and if reactions mediated by the high temperatures available at the beginning of each cycle give rise to the excess of nuclear particles over antiparticles that is observed today, one can draw a strong conclusion. The evolution of the universe from the beginning of each bounce until the formation and dissolution of galaxies will follow the same course of events during each cycle. As the cycles become longer, however, the phenomena we have described for the later stages of the open universe will become important. For example, on the first cycle in which the expansion stage lasts longer than a ten-billionth of the average proton life-time (that is, longer than about 10^{20} years) the energy gained during the contraction stage will be dominated by the photons emitted in the course of proton decay. Subsequent cycles may then lengthen by a factor of 1,000 instead of a factor of about two. Even larger expansion factors, on the order of 10^{12}, might result from the effects of coalescing black holes.

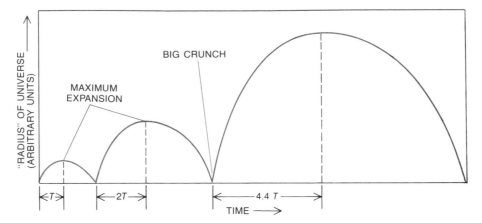

Figure 95 UNIVERSE MAY BE CYCLIC if the density of matter is sufficient to close it and if physical processes as yet unknown allow the universe to "bounce," or expand once again. Because energy is gained during the contraction phase, each new cycle expands for a longer time than the preceding one.

From a human point of view perhaps the most important question one can ask about the universe concerns the future of life and intelligence. Can the habit of thinking persist indefinitely in a universe increasingly hostile to life as it exists on the earth? Several cosmologists, including Dyson and Steven C. Frautschi of the California Institute of Technology, have begun to speculate about issues such as the energy requirements for the indefinite maintenance of life and for communication among members of a society dispersed across increasingly large regions of space.

Dyson assumes that the embodiment of life and consciousness need not be limited to cells and DNA; instead, he maintains, the essential feature of consciousness is complexity of structure, which can be realized in any convenient material. Thus he assumes that the idea of a sentient computer or a sentient cloud cannot simply be dismissed as philosophically incoherent.

Given these assumptions, the changes in the environment caused by the death and cooling of stars and their evaporation from galaxies need not be insurmountable to a system that could be counted as alive and intelligent. For example, in principle energy could be extracted from the gravitational field of a supermassive black hole.

The most efficient known scheme for extracting the energy of a black hole was put forward by Charles W. Misner of the University of Maryland, Kip S. Thorne of the California Institute of Technol-

ogy and John Archibald Wheeler of the University of Texas at Austin (see Figure 96). A rigid shell is constructed around a rotating black hole, and surplus material from the civilization is sent by a satellite close to the event horizon, the boundary of the region from which nothing can escape. Near the event horizon the material is jettisoned at the equator of the black hole. As a result both the gravitational potential energy of the jettisoned material and the rotational energy of the black hole are decreased. Because the total energy of the satellite and the black hole must be conserved, the energy of the satellite is increased by an amount equal to the two energy reductions. In principle the energy gained by the satellite could be converted into a useful form. The decay of protons and neutrons, however, may bring about a fundamental change, since it seems unlikely (although perhaps not impossible) that intelligence could be based on the tenuous foundation of electrons and positrons. Moreover, if the universe is closed, the conditions necessary for life may exist only for certain periods during each cycle.

In an open universe the ultimate impediment to life is quite different. With the evaporation of black holes there will be a cosmic energy shortage, because as the expansion of the universe continues the remaining particles and photons lose energy. Any constant rate of energy consumption by life forms will in due course become untenable. On the other hand, Dyson proposes that increasingly long periods of hibernation, during which energy is not

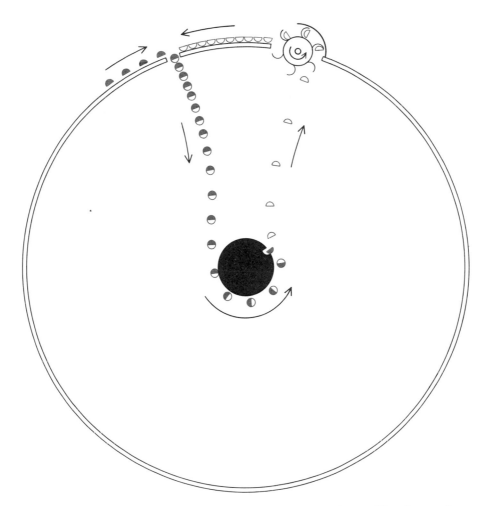

Figure 96 COSMIC ENERGY CRISIS could be delayed for a long time by extracting energy from the gravitational field of a black hole, as shown in this scheme. A rigid shell is constructed around a rotating black hole and surplus material is sent by satellite close to the event horizon, where the material is jettisoned at the equator of the black hole. In principle the energy gained by the satellite could be converted into a useful form.

consumed, could be interspersed with periods of consumption. Hence there is the potential for extraordinarily long-lived civilizations in this universe.

The starting point of contemporary cosmology, as we have emphasized, is the big-bang theory. Our quantitative treatment of the evolution of the big bang into the remote future is based on the new theories that describe the physics of elementary particles. In spirit, however, these ideas were expressed by the 11th-century Persian astronomer and poet Omar Khayyám, whose poem *The Rubáiyát* was translated by the English writer Edward Fitzgerald. In one quatrain Omar wrote:

> *With Earth's first Clay They did*
> *the Last Man's knead,*
> *And then of the Last Harvest sow'd*
> *the Seed:*
> *Yea, the first Morning of Creation*
> *wrote*
> *What the Last Dawn of Reckoning*
> *shall read.*

POSTSCRIPT

The future will be largely determined by whether the universe is open or closed. If the mean density is great enough, a "big crunch" is the inevitable consequence of the "big bang." If not, the universe will continue to expand forever. A wealth of detail is added to this basic picture by continuing advances in particle physics, observational and theoretical astrophysics and cosmology. Remarkably, the outlook for the future has changed little since our article. But five points are worth discussing.

First, we assumed the physics of proton decay, addressed in Chapter 6, "The Decay of the Proton," by Steven Weinberg. However, as discussed in Chapter 7, "The Search for Proton Decay," by J. M. Fredrick Reines and Daniel Sinclair, experiments now show that the proton lifetime exceeds that predicted by many models. The experiments continue. Perhaps observation of proton decay is just around the corner (say a lifetime of 10^{33} years); perhaps it is just beyond the reach of current experiments (say 10^{34}–10^{35} years); perhaps the proton's lifetime is significantly longer (say 10^{40} years). The precise value of the proton's lifetime, within such a range, does not qualitatively change our predictions.

At the other extreme, if the proton is absolutely stable then the discussion by Dyson would obtain. There is an important case between these extremes in which the strong, weak and electromagnetic interactions do not lead to proton decay, but a quantum theory of gravity would. Here a single proton collapsing spontaneously forms a black hole with the proton's mass, charge and spin. In some models, every proton should do this eventually. Once this happened, the black hole would evaporate quickly, emitting a positron. Estimates for the time of this mechanism are anywhere from 10^{44} to 10^{150} years. The sequence of events we discussed for the unstable proton would still hold, albeit with considerably expanded time scales.

Second, we still do not know what form is taken by most of the matter in the universe. Our predictions for an indefinitely expanding universe were based on interactions of only the electrons and positrons that remain after proton decay. We did not discuss what significant interactions could occur over long times if the universe is dominated by axions, massive neutrinos, supersymmetric particles or any of the other possibilities discussed in Chapter 1, "Dark Matter in the Universe," by Lawrence Krauss. For example, if galaxies and their clusters are dominated by massive weakly interacting particles that are unstable and decay to lighter weakly interacting particles, the decay might occur in a time approximately that of the current age of the universe. The weakly interacting decay products would now dominate the mass of the universe. In an expanding universe collisions between these particles would occur infrequently enough so that only their decays would affect the future of the universe. If the decay products are light enough, they will eventually lose energy by the Hubble expansion red shift. When the universe is ten times or so its current age, protons will dominate the energy of the universe.

Third, the past few years have seen significant advances in our knowledge of the present structure of the universe. Recent radio observations yield better limits on the anisotropy of the cosmic background of microwave radiation. Detection of an anisotropy would show the variation of the universe's properties with direction. After compensating for the velocity of the earth and our galaxy with respect to the "average," these limits are consistent with the minimum values allowed by current theories and the approximations in our article.

Fourth, a fascinating result comes from the observations of the locations of galaxies. A careful survey of a sky section out to about one-third of a billion light-years shows that galaxies appear to be concentrated on two-dimensional surfaces instead of being spread uniformly throughout the volume of space. On this scale, our universe looks like a cosmic honeycomb, or a cluster of soap bubbles, rather than a homogeneous medium. How long this structure will persist and its eventual impact on the future of the universe are unknown.

Finally, there has been no new information on the most important question for the future of the universe—whether the universe is open, closed or critical. As described in Chapter 11, "The Inflationary Universe," by Alan Guth and Paul Steinhardt, an inflationary scenario gives a natural, even compelling to some, explanation of why the density of matter and energy should be extraordinarily close to the critical density. It does not, however, give any guide as to whether the ratio of the density to the critical density is slightly larger (eventual contraction), exactly equal (expansion at a continually decreasing rate) or slightly less (eventual expansion at a constant rate). This fundamental question is likely to be extraordinarily difficult to answer because the density, if the inflationary scenario is correct, will be extremely close to the critical density whether the universe is open or closed.

In summary, the slow changes in the gross structure of the universe that predict the cosmological future are only meaningful over very long times. The microwave background will not cool and the galaxies will not dissipate in our lifetimes. Nevertheless we can push the concepts of particle physics and cosmology to their logical conclusion and thus glimpse the far future.

The Authors

LAWRENCE M. KRAUSS ("Dark Matter in the Universe") holds a joint appointment in the departments of physics and astronomy at Yale University and the Harvard/Smithsonian Center for Astrophysics. He was an undergraduate student at Carleton University in Ottawa and did his graduate work at the Massachusetts Institute of Technology, where he obtained his Ph.D. in 1982. He was a junior fellow at Harvard University before taking his current position at Yale in 1985. He is the author of the forthcoming book *The Fifth Essence: The Search for Dark Matter in the Universe* (Basic Books, scheduled for 1989).

JOHN D. BARROW and **JOSEPH SILK** ("The Structure of the Early Universe") are theoretical astrophysicists with a special interest in cosmology. Barrow is at the University of Oxford. He studied mathematics as an undergraduate at the University of Durham and obtained his doctorate in cosmology from Oxford in 1977. After spending a year as Lindemann fellow at the University of California at Berkeley, he returned to Oxford to become a faculty member in the department of astrophysics. Silk is professor of astronomy at Berkeley. He studied mathematics at the University of Cambridge, received his Ph.D. in astronomy from Harvard University and spent postdoctoral years at the Institute of Astronomy at Cambridge and at Princeton University. Silk is the author of *The Big Bang: The Creation and Evolution of the Universe* (W. H. Freeman and Company, 1988). He and Barrow are coauthors of *The Left Hand of Creation* (Basic Books, 1983).

JOSEPH SILK, ALEXANDER S. SZALAY and **YAKOV B. ZEL'DOVICH** ("The Large-Scale Structure of the Universe") are astrophysicists. Szalay is in the department of physics at Eötvös University in Budapest, where he obtained his Ph.D. in astrophysics in 1975. After periods of postdoctoral study at Berkeley, the University of Chicago and Moscow University, he returned to Eötvös in 1980. Szalay writes that his interest in cosmology was inspired by Zel'dovich, formerly professor of astrophysics at Moscow University and director of the theoretical group at the Institute of Physical Problems in Moscow. Zel'dovich was a member of the Soviet Academy of Sciences and a foreign member of both the Royal Society of London and the U.S. National Academy of Sciences. Zel'dovich died in 1987. Notes on Silk appear in the preceding biography.

HOWARD GEORGI ("A Unified Theory of Elementary Particles and Forces") is professor of physics at Harvard University. A graduate of Harvard College, he received his Ph.D. from Yale University in 1971. He then returned to Harvard, where he held a succession of postdoctoral fellowships before joining the faculty in 1976. In recent years he has published numerous papers, many in collaboration with other leading physicists, on aspects of the unification of the laws of physics.

GERARD 'T HOOFT ("Gauge Theories of the Forces between Elementary Particles") is professor of physics at the University of Utrecht, from which he received his bachelor's degree in 1969 and his Ph.D. in 1972 and to which he returned as a member of the faculty after two years at the European Organization for Nuclear Research (CERN). "My interest in physics dates back to the age of six," he writes. "I thought I was going to choose between painting, music and physics. I chose physics, inspired by other physicists in my family, but I still like to make a painting or play the piano every now and then."

STEVEN WEINBERG ("The Decay of the Proton") is Josey Professor of Science at the University of Texas at Austin. He did his undergraduate work at Cornell University, studied for a year at the Niels Bohr Institute in Copenhagen and received his Ph.D. from Princeton University in 1957. Thereafter he worked at Columbia University, the Lawrence Radiation Laboratory of the University of California, the University of California at Berkeley, the Massachusetts Institute of Technology and Harvard University. He is the author of *Gravitation and Cosmology: Principles and Applications of the General Theory of Relativity* (Wiley, 1972), *The First Three Minutes: A Modern View of the Origin of the Universe* (Basic Books, 1988) and *The Discovery of Subatomic Particles* (Scientific American Library, 1984). He has received numerous honors for his work on the theory of elementary particles, including nine honorary degrees and the 1979 Nobel Prize in physics, which he shared with Sheldon Lee Glashow and Abdus Salam. Weinberg is a member of the National Academy of Sciences and the Royal Society of London.

J. M. LOSECCO, FREDERICK REINES and **DANIEL SINCLAIR** ("The Search for Proton Decay") are experimental physicists. LoSecco is in the department of physics at the University of Notre Dame. He is a graduate of the Cooper Union School of Engineering and Science and of Harvard University, where he received his Ph.D. in 1976. He did research at Harvard, the University of Michigan and the California Institute of Technology, concentrating on the behavior and properties of the neutrino as well as on the decay of the proton, before moving to Notre Dame in 1985. Reines is professor emeritus of physics at the University of California at Irvine. He is a graduate of the Stevens Institute of Technology and New York University, where he received his Ph.D. in 1944. In 1959, after fifteen years on the scientific staff of the Los Alamos National Laboratory, he joined the faculty of the Case Institute of Technology. In 1966 he moved to Irvine, where he was the founding dean of the school of physical sciences. For his work, which includes studies of muons, neutrino scattering and proton stability, Reines received the J. Robert Oppenheimer Memorial Prize in 1981 and the National Medal of Science in 1985. Sinclair is professor of physics at the University of Michigan. He studied at the University of Glasgow, which awarded him a Ph.D. in 1957 for research on the properties of pions. Since then he has taught at the University of Michigan.

RICHARD A. CARRIGAN, JR., and **W. PETER TROWER** ("Superheavy Magnetic Monopoles") are experimental physicists who are both interested in the experimental evidence concerning the existence of monopoles. Carrigan's B.S. (1953), M.S. (1956) and Ph.D in physics (1962) are all from the University of Illinois at Urbana-Champaign. He served as a research physicist at Carnegie–Mellon University from 1961 to 1964 and as assistant professor from 1964 to 1968. In 1968 he moved to the Fermi National Accelerator Laboratory, becoming assistant head of the research division in 1977. He is now head of Fermilab's Office of Research and Technology Applications. Trower joined the faculty at the Virginia Polytechnic Institute and State University in 1966 and is currently in their department of physics. His degrees are from the University of Illinois: B.S. (1957), M.S. (1963) and Ph.D. in physics (1966). In addition to their work on magnetic monopoles both men do research on well-established elementary particles.

JAY M. PASACHOFF and **WILLIAM A. FOWLER** ("Deuterium in the Universe"). Pasachoff is director of the Hopkins Observatory and Field Memorial Professor in the department of astronomy at Williams College. He was graduated from Harvard College in 1963 and also received his master's and doctorate degrees at Harvard. Fowler is Institute Professor of Physics Emeritus at the California Institute of Technology. He obtained his bachelor's degree in engineering physics at Ohio State University in 1933 and then went to Caltech as a graduate student and received his Ph.D. in 1936. Except for periods as a Guggenheim fellow at the University of Cambridge and as a visiting lecturer at a number of institutions in the United States and Europe, he has remained at Caltech ever since. Fowler shared the 1983 Nobel Prize in physics for his work in the synthesis of the chemical elements.

FRANK WILCZEK ("The Cosmic Asymmetry between Matter and Antimatter") is a professor at the Institute for Advanced Study, Princeton. He received his B.S. in mathematics at the University of Chicago in 1970 and his Ph.D. in physics from Princeton University in 1974. He has received numerous prizes and honors, including a John and Catherine MacArthur Foundation Fellowship and the J. J. Sakurai Prize. With Betsy Devine he is the author of *Longing for the Harmonies: Themes and Variations from Modern Physics* (Norton, 1988).

ALAN H. GUTH and **PAUL J. STEINHARDT** ("The Inflationary Universe") are physicists who share an interest in the early history of the universe. Guth attended the Massachusetts Institute of Technology as an undergraduate and as a graduate student; his Ph.D. in physics was awarded by MIT in 1972. He held postdoctoral positions at Princeton University, Cornell University and the Stanford Linear Accelerator Center (SLAC). While at Cornell, Henry Tye, a fellow postdoctoral worker, persuaded Guth to join him in studying the production of magnetic monopoles in the early universe. He continued the work at SLAC, then returned to the department of physics at MIT. Steinhardt was graduated from the California Institute of Technology with a B.S. in 1974. His M.A. (1975) and Ph.D. (1978) in physics are from Harvard University. From 1979 to 1981 he was a junior fellow in the Society of Fellows at Harvard. In 1981 he moved to the University of Pennsylvania, where he is in the department of physics.

DUANE A. DICUS, JOHN R. LETAW, DORIS C. TEPLITZ and **VIGDOR L. TEPLITZ** ("The Future of the Universe") are physicists who share an interest in the relations between high-energy physics and cosmology. Dicus is a professor of physics at the University of Texas at Austin. His bachelor's and master's degrees are from the University of Washington and his doctorate in physics is from the University of California at

Los Angeles. After serving as research associate at UCLA, the Massachusetts Institute of Technology and the University of Rochester, he moved to the University of Texas in 1973. Letaw is technical director of the Severn Communications Corporation in Severna Park, Maryland. He received his B.A. at Clark University in 1975 and his Ph.D. in physics from the University of Texas at Austin in 1981. He is currently working on space radiation effects on both astronauts and microelectronics. Doris Teplitz is adjunct at the University of Maryland at College Park. Her B.A. is from Wellesley College and her Ph.D. in physics is from Northeastern University. She has compiled and edited a graduate text on electromagnetism. Vigdor Teplitz is deputy chief of the Strategic Affairs Division of the U.S. Arms Control and Disarmament Agency and adjunct professor of physics at the University of Maryland. His B.S. is from MIT and his Ph.D. in physics is from the University of Maryland. He has held appointments at MIT and Virginia Polytechnic Institute and State University. In 1988 he served as program director for theoretical physics at the National Science Foundation.

Bibliography

Comprehensive readings:

Barrow, John D., and Joseph Silk. 1983. The left hand of creation: The origin and evolution of the expanding universe. New York: Basic Books.

Close, Frank. 1983. The cosmic onion: Quarks and the nature of the universe. London: Heinemann Educational Books.

Hawking, Stephen W. 1988. A brief history of time: From the big bang to black holes. New York: Bantam.

Kippenhahn, Rudolf. 1983. 100 billion suns: The birth, life, and death of the stars. (Trans. by Jean Steinberg.) New York: Basic Books.

Pagels, Heinz R. 1982. The cosmic code: Quantum physics as the language of nature. New York: Simon and Schuster.

Silk, Joseph. 1988. The big bang. New York: W. H. Freeman and Company.

Trefil, J. S. 1980. From atoms to quarks: An introduction to the strange world of particle physics. New York: Charles Scribner's and Sons.

Weinberg, Steven. 1988. The first three minutes: A modern view of the origin of the universe. New York: Basic Books.

Wilczek, Frank and Betsy Devine. 1988. Longing for the harmonies: Themes and variations from modern physics. New York: Norton.

Zee, A. 1986. Fearful symmetry: The search for beauty in modern physics. New York: Macmillan Publication Company.

Readings from the magazine:

Abbott, Larry. 1988. The mystery of the cosmological constant. Scientific American 258 (May): 106–113.

Bloom, Elliott D., and Gary J. Feldman. 1982. Quarkonium. Scientific American 246 (May): 66–77.

Cline, David B., Carlo Rubbia and Simon van der Meer. 1982. The search for intermediate vector bosons. Scientific American 246 (March): 48–59.

Freedman, Daniel Z., and Peter van Nieuwenhuizen. 1978. Supergravity and the unification of the laws of physics. Scientific American 238 (February): 126–143.

Freedman, Daniel Z., and Peter van Nieuwenhuizen. 1985. The hidden dimensions of spacetime. Scientific American 252 (March): 74–81.

Glashow, Sheldon L. 1975. Quarks with color and flavor. Scientific American 233 (October): 38–50.

Green, Michael. 1986. Superstrings. Scientific American 255 (September): 48–60.

Haber, Howard E., and Gordon L. Kane. 1986. Is nature supersymmetric? Scientific American 254 (June): 52–60.

Harari, Haim. 1983. The structure of quarks and leptons. Scientific American 248 (April): 56–68.

Jackson, David J., Maury Tigner and Stanley Wojcicki. 1986. The superconducting supercollider. Scientific American 254 (March): 66–77.

Lederman, Leon M. 1978. The upislon particle. Scientific American 239 (October): 72–80.

Mistry, Nariman B., Ronald A. Poling and Edward H. Thorndike. 1983. Particles with naked beauty. Scientific American 249 (July): 106–115.

Parker, E. N. 1983. Magnetic fields in the cosmos. Scientific American 249 (August): 44–54.

Perl, Martin L., and William T. Kirk. 1978. Heavy Leptons. Scientific American 238 (March): 50–57.

Quigg, Chris. 1985. Elementary particles and forces. Scientific American 252 (April): 84–95.

Rebbi, Claudio. 1983. The lattice theory of quark confinement. Scientific American 248 (February): 54–65.

Rubin, Vera C. 1983. Dark matter in spiral galaxies. Scientific American 248 (June): 96–108.

Schwitters, Roy F. 1977. Fundamental particles with charm. Scientific American 237 (October): 56–70.

Weinberg, Steven. 1974. Unified theories of elementary-particle interaction. Scientific American 231 (July): 50–59.

Wilson, Robert R. 1980. The next generation of particle accelerators. Scientific American 242 (January): 42–57.

Original literature for the advanced reader:

Annual review of astronomy and astrophysics. Palo Alto, Calif.: Annual Reviews.

Annual review of nuclear and particle science. Palo Alto, Calif.: Annual Reviews.

INDEX

Page numbers in *italics* indicate illustrations.